The New Edition

TECHNICAL ENGLISH at WORK

David Clarke

Cornelsen

Technical English at Work – The new edition wurde geplant und entwickelt von der Redaktion Moderne Fremdsprachen des Cornelsen Verlags, Berlin.

Verfasser:	David Clarke, Witten
Beratung:	Wilfried Jörres, Viersen
	Pano Papaspyrou, Soest
	Thomas Pache, Euskirchen
	Dieter Rackelmann, Krefeld
	Sybille Weiß, Willich
Projektleitung:	Jim Austin
Außenredaktion:	Roderick A.K. Cox, Freudenstadt
Redaktionelle Mitarbeit:	Lauren Chexal, Rosalind Higman,
	Fritz Preuß (Wörterverzeichnisse)
Bildredaktion:	Gertha Maly
Layout und technische Umsetzung:	Klein & Halm Grafikdesign, Berlin
Umschlaggestaltung:	Ellen Meister
Titelfoto:	Fotosearch RF
Illustrationen:	Oxford Designers & Illustrators

Erhältlich sind auch:

Audio-CDs	ISBN 978-3-06-020394-9
Handreichungen für den Unterricht	ISBN 978-3-06-020395-6
Job Matters Elektrotechnik	ISBN 978-3-06-020401-4
Job Matters Metalltechnik	ISBN 978-3-06-020400-7

www.cornelsen.de

1. Auflage, 3. Druck 2012

Alle Drucke dieser Auflage sind inhaltlich unverändert und können im Unterricht nebeneinander verwendet werden.

Druck: Stürtz GmbH, Würzburg

ISBN 978-3-06-020393-2

Inhalt gedruckt auf säurefreiem Papier aus nachhaltiger Forstwirtschaft.

Technical English at Work – The new edition ist ein praxisnahes und handlungsorientiertes Lehrwerk für die Fachschule, für die innerbetriebliche Weiterbildung und für den allgemeinen Englischunterricht mit technischem Schwerpunkt.

Aufbau des Gesamtprogramms

Das Programm des vorliegenden Kursbuchs ist für die Aus- und Weiterbildung in allen technischen Berufen geeignet. Es enthält Komponenten, um die Sprachfertigkeit sowohl zu aktivieren wie auch zu schulen, um die bereits vorhandenen Englischkenntnisse aufzufrischen und – vor allem in den hinteren Units – beides anhand von fachübergreifenden Themen zu erweitern.

Darüber hinaus werden in den Begleitheften (course modules) fachspezifische Themen aus den Bereichen Elektrotechnik und Metalltechnik behandelt.

Aufbau des Lehrwerkes

Das Kursbuch beinhaltet insgesamt 16 Units, die in zwei Teile gegliedert sind:

- **Refresher Course: Units 1–6**
- **Main Course: Units 1–10**

Im vorbereitenden Refresher Course werden Grundkenntnisse wie die Wortstellung, die Pluralbildung und die Pronomen sowie Sprechfunktionen, wie *Sich vorstellen* und *Über Beruf und Hobbys* reden, aufgefrischt.

In den Units 1–3 des Main Course werden berufsbezogene Themen wie *Finding a job* und *Moving to Dublin* anhand der Geschichte eines deutschen Technikers behandelt.

Die Units 4–10 sind in zwei Teile, A und B, gegliedert. Teil A führt die Themen aus dem Berufsleben, wie *Arranging and taking part in meetings* und *Talking about jobs*, weiter und greift Themen aus dem Umfeld der Technik, wie *Global warming* oder *Future lighting technology*, auf. Im Zentrum von Teil B stehen technikspezifische Themen wie *Shapes* und *Materials*.

Die Grammatik wird in **Technical English at Work – The new edition** auf das Wesentliche begrenzt. Wichtige Strukturen werden in den Units kurz präsentiert und durch fakultative Übungen in den *Practising grammar*-Seiten am Ende jeder Unit gefestigt.

Das Kursbuch schließt mit einem Anhang, einschließlich einer Begleitgrammatik auf Deutsch und einer chronologischen sowie alphabetischen Wörterliste.

Der Verlag, die Berater und der Autor wünschen Ihnen viel Erfolg und auch Freude bei Ihrer Arbeit mit **Technical English at Work – The new edition**.

Tr00 Text bzw. Dialog auf CD   Partnerarbeit   Gruppenarbeit

Appendix

Getting to know each other

1 Use the speech bubbles and the material in the box to introduce yourself.

Alan Garner is a teacher at a language school in Hastings.
He's with a group of new students on a technical English course.

Hello, everybody. My name's Alan Garner. I'm English. I'm from Hastings. I'm a teacher of technical English at a language school.

Hello. I'm Alara Tarik. I'm Turkish. I'm from Cologne. I'm an electrician at a power plant.

Hello. My name's Paul Schäfer. I'm from Cologne, too. I'm a maintenance technician at a furniture factory.

Hello. I'm Alain Durand. I'm French and I'm from Rouen in Normandy. I'm a self-employed heating engineer.

Hi, everybody. My name's Halina Piatkowski. I'm unemployed at the moment, but I'm on a retraining programme as a trainee computer technician.

Good morning, everybody. My name's Anna Strobl and I'm Austrian. I'm from Linz. I'm a chemist in a food factory.

Afghan	*Afghanistan*	German	*Germany*	Portuguese	*Portugal*
Albanian	*Albania*	Greek	*Greece*	Russian	*Russia*
Austrian	*Austria*	Hungarian	*Hungary*	Serbian	*Serbia*
British	*Britain*	Italian	*Italy*	Slovak	*Slovakia*
Croatian	*Croatia*	Kazach	*Kazachstan*	Spanish	*Spain*
Czech	*Czech Republic*	Moroccan	*Morocco*	Turkish	*Turkey*
Dutch	*the Netherlands*	Polish	*Poland*	Ukrainian	*Ukraine*

a chemist • a computer technician • a draughtsperson • an electrician • an apprentice electrician • an engineer • an electronics technician • a heating engineer • a lab assistant • a lathe operator • a machine operator • a maintenance technician • an industrial / a construction mechanic • a mechanic • a technical assistant • a telephone engineer • ...

2 Use the speech bubbles and the material in the box to ask somebody their name and how to spell it.

Can you tell me your name, please?

Sure. It's Halina Piatkowski.

Can you spell that for me, Halina?

Of course. It's P – I – A – T – K – O – W – S – K – I.

Sorry, was that K – I at the end?

That's right. Kilo – India.

Thanks. I have that now.

How do you spell your surname, Paul?

It's S – C – H – A with an umlaut – F – E – R.

Thanks. And how about you, Anna? Is it 'Strobel' with E – L?

No. Just L. It's S – T – R – O – B – L.

Thanks a lot.

Aa	Alpha	**Gg**	Golf	**Mm**	Mike	**Ss**	Sierra	**Yy**	Yankee
Bb	Bravo	**Hh**	Hotel	**Nn**	November	**Tt**	Tango	**Zz**	Zulu
Cc	Charlie	**Ii**	India	**Oo**	Oscar	**Uu**	Uniform		
Dd	Delta	**Jj**	Juliette	**Pp**	Papa	**Vv**	Victor		
Ee	Echo	**Kk**	Kilo	**Qq**	Quebec	**Ww**	Whiskey		
Ff	Foxtrot	**Ll**	Lima	**Rr**	Romeo	**Xx**	X-ray		

3 Link the questions 1–10 to the answers A–J.

1 I'm from Austria. How about you?
2 Do you work here in London?
3 What's your job?
4 Do you mind if I smoke?
5 Is your name Baldo?
6 Is this your first visit to London?
7 May I get you another drink?
8 You're with Ruhrmetall, aren't you?
9 What are your hobbies?
10 Are you doing anything later?

A I'm a computer technician. What about you?
B No, it isn't. I'm Branco. Branco Malik from Zagreb.
C I'm French. I live in Lyon.
D I have two – motorcycling and my dog.
E No, it isn't. I live here, actually.
F No, I'm not. I'm free this evening.
G No, I don't. I work in Manchester.
H Thanks. I'd like a beer, please.
I No, I don't. Go ahead.
J Well, it's called Rheinmetall. I'm an engineer.

4 Use the dialogue guide to talk to one another.

A Where do you live? / work?

B I live / work in ... It's near ... / part of ...

A What's your job?

B I'm a / an
- chemist.
- computer technician.
- engineer.
- apprentice mechanic.
- ...

A What are your hobbies?

B My big hobby is
- motorcycling.
- car mechanics.
- football/basketball.
- my local carnival society.
- working out at the gym.
- playing in a band.
- ...

A What do you do in your spare time?

B I enjoy / like / spend time
- surfing the net.
- meeting my friends.
- shopping.
- LAN parties.
- visiting chat rooms.
- playing online poker.
- playing video games.
- watching TV.
- chilling out.
- internet telephoning.
- dancing.
- ...

A Are you free
- tonight?
- tomorrow?
- Friday/... evening?
- at the weekend?

B
- Yes, I am.
- Sorry, I'm afraid I'm busy then.

A Why don't we
- meet for a beer?
- go to the movies/...?
- go shopping?
- ...?

B
- OK.
- Sorry, I can't.

Numbers and letters

We live in a world of numbers, mostly **identification** and **registration numbers**. You can see fourteen examples below.

5 **a) Use the box to give the English terms for the items.**

account number	invoice number	phone number
bar code	lottery numbers	postal code
car number	membership number	road number
credit card number	order number	ticket number
email address	passport number	

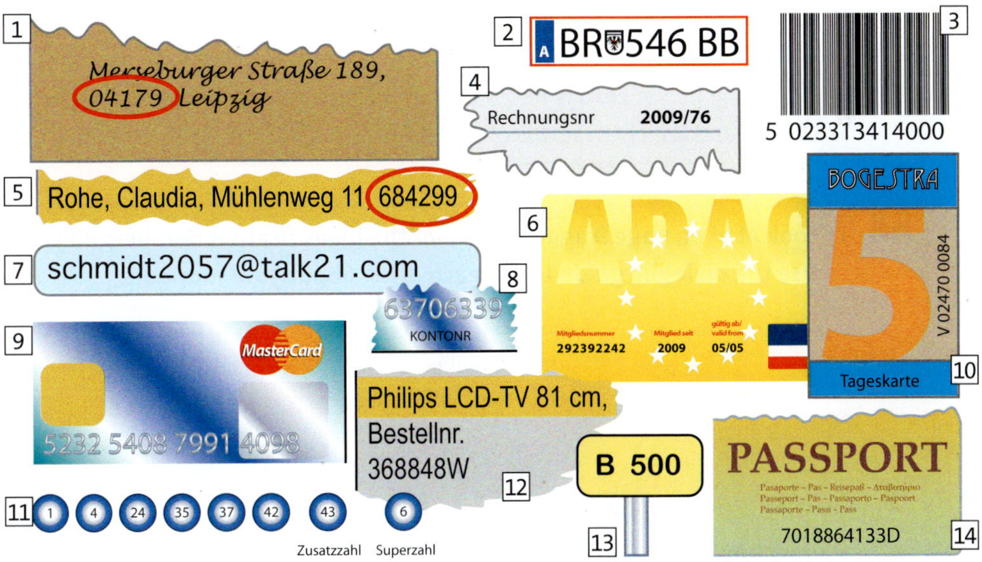

– I'm sure number … – I think number …	is is probably could be	a/an …

b) Work with a partner. Dictate the numbers in a) to each other. Read the information in the infobox first.

Saying big numbers

Say the numbers – also double numbers, e.g. **33** – one after the other.
For a nought (**0**), use either the letter **O** (oh) or *zero* (**0**):

We write:	**We say:**
18995001	one – eight – nine – nine – five – oh/zero – oh/zero - one

German "Null" in English

There are several ways of expressing "Null" in English:

How many **noughts** (BE) / **zeros** (AE) are there in a million?	mathematical numbers
We say data file names with **nought** (BE) or **zero** (BE/AE).	IT applications
We won three – **nil** (BE) / **zero** (AE).	score in team games
It was ten degrees below **zero** (BE/AE) last night.	temperatures
... three – two – one – **zero** (BE/AE). Liftoff!	countdowns to start
My phone number is three – four **oh** (BE) / **zero** (AE) – six.	series of numbers, e.g. phone numbers

! The use of **zero** is becoming much more common in British English, particularly in series of numbers and all technical contexts.

c) Look at a) on page 9 again. Which of the identification numbers do you have? Use the dialogue guide to ask each other about your numbers.

A	What's your	mobile phone car/motorbike ...	number?
B	It's ... I don't have a How about you?	 car. credit card. passport.	
A	I have	an account number, a phone number, ... and ...	
	My	postcode car/motorbike number ...	is ...

6 Tr 2

You work for Kent Automotive Limited in Ashford. Listen to the answerphone message and note the two identification numbers and the two phone numbers.

Mathematical terms and calculations

Basic mathematical terms					
+	plus	÷	divided by	5^3	five cubed
–	minus	=	is/equals	2^4	two to the power of 4
×	multiplied by	4^2	four squared	π (pi)	approx 3.14

	We write:	**We say:**
Examples:	$2 + 3 = 5$	Two plus three is/equals five.
	$9 - 6 = 3$	Nine minus six is/equals three.
	$3 \times 3 = 9$	Three multiplied by three is/equals nine.
	$8 \div 2 = 4$	Eight divided by two is/equals four.
	$5^3 = 125$	Five cubed is/equals one hundred and twenty-five.
	$4^2 \times 2^4 = 256$	Four squared multiplied by two to the power of 4 is/equals two hundred and fifty-six.
	$\pi \times 3^2 = 3.14 \times 9 = 28.26$ sq cm	(The area of the circle) ... is twenty-eight point two six square centimetres.

7 **First work out the answers to the calculations, then dictate the calculations to a partner. Change roles. Then think up two more calculations using language from the box above.**

1 $19 \times 8 \div 4 = 38$
2 $195 + 205 \times 4 \div 20 =$ ...
3 $42 + 33 =$ ...
4 $3^2 + 4^3 =$ ...
5 $3^4 \times 2 =$ ...
6 $6 \times 7 - 4 =$ ...
7 $24 \div 6 + 2 =$ ...
8 $101 - 11 + 3 =$ ...
9 $2^3 + 7 =$ ...

8 **Use the labelled diagram and words from the box to describe a football pitch.**

A standard football pitch has a length of between ... and ... metres, and a width ranging from ... to ... metres. It has an area of between ... and ... square metres, depending on its length and width. The playing area is divided into ... of equal size. ...

pitch, playing area • length, width, area (length x width) • corner, corner post • sideline, goalline • half/halves • goal, goalpost, crossbar, goal net • team, opposing teams, field player, goalkeeper

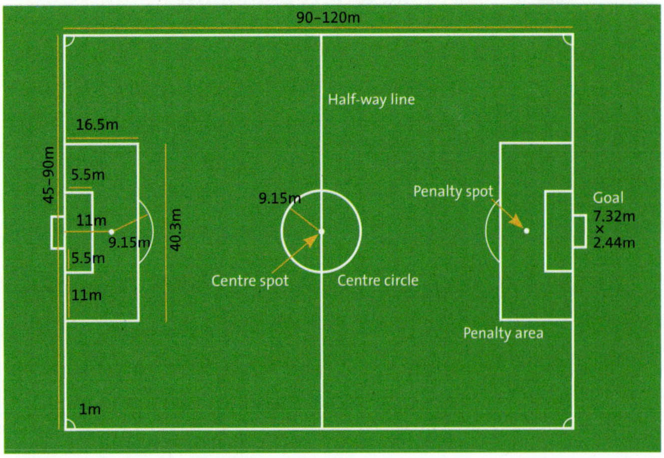

Personalpronomen

- Um eine eintönige Wiederholung von Nomen und Namen zu vermeiden, benutzt man **Pronomen**.
- Hier ist eine **Übersicht** der Personalpronomen mit den deutschen Entsprechungen.

Subjektform		Objektform		
		Wenfall		Wemfall
I	*ich*	me	*mich*	*mir*
you	*du/Sie*	you	*dich/Sie*	*dir/Ihnen*
he	*er*	him	*ihn*	*ihm*
she	*sie*	her	*sie*	*ihr*
it	*er/sie/es*	it	*es/ihn/sie*	*ihm/ihr*
we	*wir*	us	*uns*	*uns*
you	*ihr/Sie*	you	*euch/Sie*	*euch/Sie*
they	*sie*	them	*sie*	*ihnen*

!

Im Englischen verwendet man immer *it* für Sachen,
(*he/she* wird nur für Personen und Tiere mit Namen benutzt):

Der Bus war verspätet. ***Er*** *hatte eine Panne.* — *The bus was late.* ***It*** *had a breakdown.*
Die Uhr geht nach. ***Sie*** *ist kaputt.* — *The clock is slow.* ***It*** *is broken.*

1 **Replace the highlighted nouns and/or names with the correct pronoun.**

1 The students are in school now.
2 Alain is from Rouen in France.
3 The hammer is on the workbench.
4 Alara is an electrician.
5 Paul and Alara are from Cologne.
6 Alan's car is in the car park.
7 Paul has two sisters.
8 Halina is a trainee computer technician from Gdansk.
9 Alan and Sally are teachers at a language school.
10 The book is on the table.

2 **Fill in the correct pronouns.**

1 Paul Where are *you* [1] from, Halina?
Halina ... 'm [2] from Gdansk. ... 's [3] in Poland.
Paul ... [4] know. In German ... 's [5] Danzig.
Halina And what about ... [6], Paul? Where do ... [7] live?
Paul ... [8] live in Cologne.
Halina Oh, yes. That's like Alara. ... [9] lives in Cologne, too.

2 Alan Hi, Anna. ... 'd [1] like to introduce Inga and Vadim. ... 're [2] two new students.
Anna Hi. ... 's [3] nice to meet ... [4]. Where are ... [5] from?
Inga ... 're [6] from St Petersburg.
Anna St Petersburg? That's a beautiful city.
Vadim Yes, it is. Some people think ... 's [7] the most beautiful city in Europe.

Possessivbegleiter

- **Possessivbegleiter** zeigen an, wem etwas gehört.
  Sie stehen – wie ein Adjektiv – unmittelbar vor dem Nomen.

- Hier ist eine **Übersicht** der Possessivbegleiter mit Beispielsätzen:

I	Sally is **my** sister.	
you	Can you spell **your** name, please?	
he	What is **his** job?	
she	Is this **her** car?	
it	Ireland is in the EU. **Its** capital is Dublin.	! **it's** = it is

we	There are 14 students in **our** group.	
you	Please open **your** books at page 23.	
they	Four students can't find **their** homework.	! **there** = *da, dort*

3 **Translate the highlighted part of the sentences into English.**

1 Wie ist Ihr Name, bitte?
2 Ist der Golf sein Auto?
3 Anna schreibt ihrer Familie eine Ansichtskarte.
4 Die Studenten müssen ihre Bücher selbst kaufen.
5 Momentan kann ich meinen Wagen nicht benutzen. Seine Batterie ist leer.
6 Der Polizist fragte mich nach meiner Adresse.
7 Dieses Klassenzimmer ist viel zu klein für unsere Gruppe.
8 Die Schüler sagen, dass ihr Lehrer nicht erschienen sei.
9 Sie dürfen Ihr Auto hier nicht parken.
10 Markus ging hinaus, ohne seinen Test abzugeben.
11 Die junge Frau da drüben? Ihr Name ist Anna Strobl.
12 Sieh dir den armen Vogel an. Sein Bein ist verletzt.

Das Verb *be*

Bejaht			Verneint	
Langform		Kurzform	Langform	Kurzform
I am	*ich bin*	I'm	I am not	I'm not
you are	*du bist/Sie sind*	you're	you are not	you're not/you aren't
he is	*er ist*	he's	he is not	he's not/he isn't
she is	*sie ist*	she's	she is not	she's not/she isn't
it is	*es ist*	it's	it is not	it's not/it isn't
we are	*wir sind*	we're	we are not	we're not/we aren't
you are	*ihr seid/Sie sind*	you're	you are not	you're not/you aren't
they are	*sie sind*	they're	they are not	they're not/they aren't

- Im Gespräch und in E-Mails usw. benutzt man meist die Kurzform. Wie im Deutschen dient der Apostroph (') als Auslassungszeichen.

Bei verneinten Aussagen kann man zwischen zwei Kurzformen – z. B. *she's not* und *she isn't* – wählen.

- Um Fragen zu bilden, tauscht man – genau wie im Deutschen – das Subjekt und das Verb einfach aus:

Anna is from Austria.	Is Anna from Austria?
Anna kommt aus Österreich.	*Kommt Anna aus Österreich?*
Paul and Halina are in the café.	Are Paul and Halina in the café?
Paul und Halina sind im Café.	*Sind Paul und Halina im Café?*

4 Fill in the correct form of the *simple present* of *be*. Use the long form.

1 My name *is* [1] Dick Brown. I ... [2] a technician with a German firm in Southport. Southport ... [3] in north-west England. It ... [4] not far from Liverpool. But at the moment I ... [5] not in Southport. I ... [6] on a training course with my firm in Dortmund.

2 There ... [1] eight technicians on the course but only one other person, Angus Cameron, ... [2] British. Angus ... [3] not from England. He ... [4] from Glasgow in Scotland.

3 We ... [1] an international group. There ... [2] technicians on the course from six countries and the instructors ... [3] German. But Angus and I ... [4] lucky. It ... [5] an international firm and the language ... [6] English, even when we ... [7] in Germany.

5 Give the short form of the highlighted long forms.

1 I am German. I am from Essen.
2 We are not teachers. We are students.
3 You are not from Vienna, are you, Anna?
– No, I am not. I am from Linz.
4 Halina is from Gdansk. She is very nice.
5 They are not in their classroom. They are in the media room today.
6 It is not very warm today, is it? It is only ten degrees.
7 Where is Alan?
– He is not here today. He is ill.
8 You are not in Sally's group, are you?
– No, we aren't. We are in Alan's.

Das Verb *have*

I have	*ich habe*	we have	*wir haben*
you have	*du hast/Sie haben*	you have	*ihr habt/Sie haben*
he has	*er hat*	they have	*sie haben*
she has	*sie hat*		
it has	*es hat*		

6 **Translate the highlighted part of the sentences into English.**

Examples:
- Das Klassenzimmer hat zwölf Tische.
  has
- Ich habe keine Stelle. Ich bin Schüler.
  don't have a job
- Haben Sie einen Hammer, bitte?
  Do you have

1. Der Computer hat vier USB-Anschlüsse.
2. Wir haben zehn Laptops vorrätig.
3. Der Computer hat keine deutsche Betriebsanleitung.
4. Die Studenten haben ihre Englischbücher dabei.
5. Ihr habt einen DVD-Recorder in eurem Klassenzimmer.
6. Diese Taschenlampe hat keine Batterie.
7. Ich habe einen Wochenendjob.
8. Paul hat ein neues Auto.

Die Wortstellung in Aussagen

	Subjekt	Verb	Objekt Wemfall	Objekt Wenfall	*andere* Satzteile
1	*Alan*	*teaches*	–	*English*	*at a school in Hastings.*
2	*Alan*	*told*	*the students*	*his name.*	
3	*Alara Tarik*	*is*	–	–	*from Cologne.*

- Im Englischen ist die Grundwortstellung in Aussagen: **Subjekt – Verb – Objekt – andere Satzteile.** [1]
- Gibt es ein **indirektes Objekt** (wem?) im Satz, steht es unmittelbar vor dem **direkten Objekt** (wen/was?). [2]
- Kommt **kein Objekt** im Satz vor, ist die Reihenfolge: **Subjekt – Verb – andere Satzteile.** [3]

7 **Put the parts of the sentences into the correct order.**

1. from Linz/is/Anna Strobl
   Anna Strobl is from Linz.
2. a new car/Paul/has
3. for their English lesson/late/were/three students
4. is/Anna/from Linz/a food chemist
5. Alan Garner/at a language school/a teacher/is/in Hastings
6. at the moment/unemployed/is/Halina
7. at a furniture factory/is/Markus/a maintenance technician
8. on the workbench/the tools/are
9. do/have/they/a big problem
10. Markus/a new CD/has

1 **Use the dialogue guide and the list of tools to ask each other about the English terms.**

A What's Steckschlüssel in English?
B It's *socket wrench*.
A Thanks.

A What's ... in English?
B It's ...

1 a socket wrench
2 a Phillips screwdriver
3 a bladed screwdriver
4 **a pair of** pliers
5 **a pair of** pincers
6 an Allen key
7 a hammer
8 a rule
9 a digital measuring slide
10 **a pair of** wire strippers
11 a round file
12 a triangular file
13 a soldering iron
14 an adjustable spanner
15 a centre punch
16 an electric drill
17 a hacksaw
18 **a pair of** metal cutters
19 a chisel
20 a spirit level
21 an electric jig saw
22 a battery screwdriver
23 a belt sander
24 an angle grinder
25 screws
26 nails

2 **First look at the examples below. Then complete the speech bubbles.**

This is a hammer.

These are screwdrivers.

Those are files.

That is a pair of metal cutters.

3 **Choose the correct form, a), b) or c).**

	Bejaht	Verneint	Frageform
Singular	*There is / There's*	*There is not / There isn't*	*Is there?*
Plural	*There are*	*There are not / There aren't*	*Are there?*

1 a) It is
b) There are — six people at the bus stop.
c) There is

2 a) There aren't
b) There is — a dog in the car.
c) This is

3 Excuse me, please. a) Are there
b) Is it — a post office near here?
c) Is there

4 a) Is there
b) There are — some books in the living room.
c) There is

5 There a) are
b) aren't — a computer in our classroom.
c) isn't

6 a) Are there
b) Is there — any shops in this road?
c) There are

4 **Look at the sketch on page 16. Then complete the sentences with *there is/are* and the missing adverbs of place (*Ortsadverbien*).**

1. There ... a pair of metal cutters on the workbench.
2. There ... three electric tools on the shelf.
3. There ... two screwdrivers in the drawer.
4. There ... a hammer on the workbench.
5. There ... a pair of pliers next to the screwdrivers.
6. There ... a hacksaw behind the socket wrench.
7. ... a rule next to the Allen key.
8. ... some screws ...
9. ... a chisel ...
10. ... two soldering irons ...
11. ... a pair of wire strippers ... the digital measuring slide.
12. ... an angle grinder ...
13. ... some nails ...
14. ... a belt sander ...

5 First, link the tools to their functions. Then use the dialogue guide to ask and answer questions about them.

1 hammer
2 screwdriver
3 chisel
4 rule
5 a pair of wire strippers
6 hacksaw
7 a digital measuring slide
8 a pair of pincers
9 a pair of metal cutters
10 soldering iron
11 round file
12 flat file

(to) cut	*schneiden*
(to) drive in	*eindrehen*
(to) join	*verbinden*
(to) knock in	*einschlagen*
(to) measure	*messen*
(to) remove	*entfernen*
(to) shape	*formen*
(to) smooth	*glatt schleifen*

a (to) cut thicker metal
b (to) cut tin or thin metal sheet
c (to) cut wire
d (to) drive in screws
e (to) join wire securely
f (to) knock in nails
g (to) measure length and distance
h (to) measure exact thickness
i (to) remove insulation
j (to) shape wood
k (to) smooth the inside of holes in metal
l (to) smooth the surface of metal

– What do you use a	hammer round file soldering iron …	for?
– You use it to	knock in nails. smooth the inside of holes in metal. join wire securely. …	
– What tool do I need to	cut a thicker piece of metal? smooth the surface of metal? measure the exact thickness of this pipe? …	
– You need	a hacksaw. a flat file. a digital measuring slide. …	

This/these, that/those

this	Singular	*diese/r/s*	**This** hammer **is** new.	*in der Nähe des Sprechers*
these	Plural	*diese*	**These** screwdrivers **are** new, too.	
that	Singular	*jene/r/s*	**That** chisel **is** blunt (*stumpf*).	*weiter weg vom Sprecher*
those	Plural	*jene*	**Those are** Jack's spanners.	

1 **Use the prompts to fill in *this/these, that/those*.**

here: *hier* (*in meiner Nähe*)
over there: *dort drüben*

1 … DVD on the shelf over there is very good.
2 … CDs here are Kate's.
3 … book here is in German.
4 … hammer here isn't heavy enough.
5 … tools over there are all broken.
6 … digital measuring slide you have in the van is inaccurate (*ungenau*).
7 … nails I have are too short. Look.
8 You need a Phillips screwdriver for … screws you have over there.

Nomen

Der Plural

Singular	Plural	Form
a chisel	two chisel**s**	Anhängen von -**s** ohne weitere Änderung
a screwdriver	three screwdriver**s**	
an electric drill	four electric drill**s**	
a centre pun**ch** a bu**s** a paint bru**sh** (*Pinsel*) a bo**x**	two centre punch**es** three bus**es** four paint brush**es** four box**es**	Anhängen von -**es** nach -**ch**, -**s**, -**sh** und -**x**
a facto**ry**	two factor**ies**	Das -**y** (**nicht** jedoch -**ey**) wird zum -**i**. Dann wird -**es** angehängt.
a kni**fe**	two kni**ves**	Endet der Singular auf -**fe**, wird das **f** zum **v**. Dann wird -**s** angehängt.
a child a foot an aircraft	three **children** two **feet** ten **aircraft**	Eine geringe Zahl von Nomen ist **unregelmäßig**. Weitere wichtige Beispiele sind: man – **men**, woman – **women** ['wɪmɪn], tooth – **teeth**

Paarwörter

Kleidung (*jeans, shorts*) und **Werkzeuge** (*pliers, metal cutters*) aus zwei miteinander verbundenen Teilen nennt man „Paarwörter" (*pair nouns*).

Beispiele	Regeldarstellung
The pliers are on the workbench. *Die Zange ist auf der Werkbank.*	Paarwörter stehen im **Plural** und sind **nicht zählbar**. Sie werden beispielsweise mit *the, these/those* und *some* (einige) benutzt.
Here **is a pair of pliers**. *Hier ist eine Zange.* **Two pairs of metal cutters** are broken. *Zwei Blechscheren sind kaputt.*	Möchte man Paarwörter **im Singular** benutzen oder **zählbar** machen , fügt man *a/... pair(s) of* hinzu.

2 **Give the plural forms of the nouns. They all have to do with cars.**

1 car radio
2 battery
3 car body (*Karosserie*)
4 car crash
5 clutch (*Kupplung*)
6 cylinder
7 fuel mix
8 light switch
9 airbag

3 **Translate the highlighted parts of the sentences into English. They are all pair nouns.**

1 Jack hat eine Zange in der Hand.
2 Ich habe drei Abisolierzangen.
3 Deine Brille ist schmutzig.
4 Diese Kneifzange ist alt.
5 Die Schere ist viel zu groß.
6 Kate hat acht Jeans!

Some/any

some	*einige*	**Some** students **are** late.	*in bejahten Aussagen*
any	*kein/e*	There **aren't any** tools in the drawer.	*in verneinten Aussagen*
	irgend-welche	**Are there any** hammers on the workbench?	*in Fragen*

■ Diese Regeln gelten auch für die Zusammensetzungen von *some* und *any*: *somebody/anybody, someone/anyone; something/anything; somewhere/anywhere.*

4 **Complete the sentences with *some, any* or one of their compounds.**

1 You can't get ... alcohol in the canteen but they have ... non-alcoholic drinks.
2 Do you have ... money with you?
– No, but I'm sure Ben has ... . Ask him.
3 Maybe Kate can give you ... help. Bob can't. He doesn't know ... about computers.
4 I can't find a rule ... . Can I have yours?
– No. You never give ... back, that's the problem.

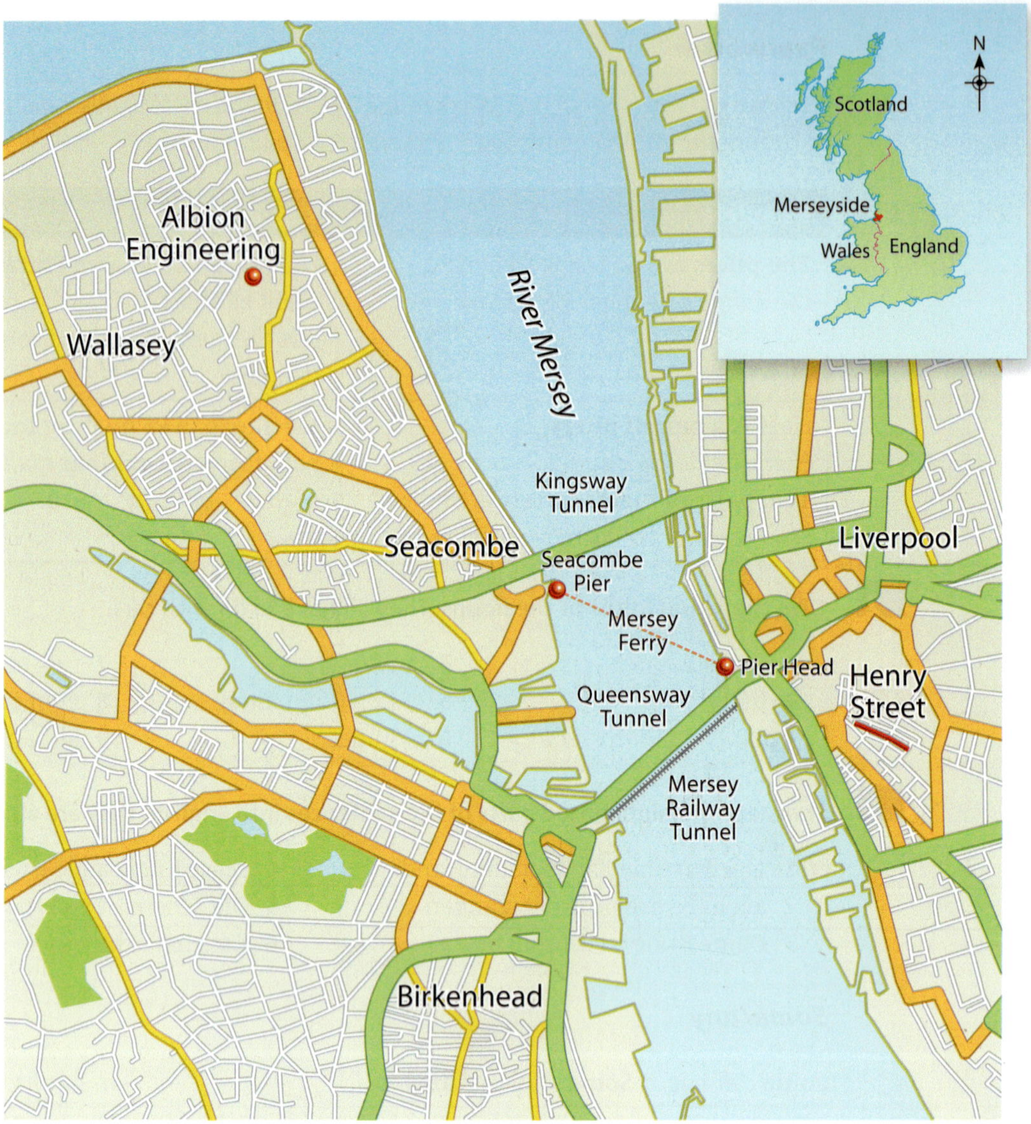

1 Look at the maps. Use the dialogue guide to describe and talk about them.

The	big small	map	shows	the	mouth of the Mersey. location of the Mersey in north-west England.

Liverpool Seacombe Birkenhead Wallasey	is	west east	of the river.
There	are is	two tunnels a ferry	between ... and ...

The Mersey Ferry runs from the ... in Liverpool to ... on the ... bank.

Tr3 Nick's day

Nick Kelly is a maintenance technician with Albion Engineering Limited in Wallasey. The firm makes automotive parts for car companies all over the world. Its biggest customers are Opel in Germany and Peugeot in France.

But Nick doesn't live in Wallasey – it's too expensive. He and his partner, Jenny, live on the opposite side of the Mersey in Henry Street. This is a street of workers' houses in the old part of Liverpool near the docks.

Albion Engineering works two eight-hour shifts, Monday to Friday. The early shift is from 6 a.m. to 2 p.m., and the late shift is from 2 p.m. to 10 p.m. Nick always does the early shift, but he doesn't mind this, even in winter. Nick doesn't find it difficult to get up, and he likes the free afternoons and evenings – and he likes the 15% shift allowance, too.

What does Nick's working day look like?

On a typical working day, Nick gets up at 5.15. First, he turns on the coffee maker and then he showers and gets dressed. When he's ready, the coffee's ready, too. He drinks two cups of coffee, but he doesn't have breakfast.

Nick leaves the house at 5.40 and drives through the Kingsway Road Tunnel to Dock Road in Wallasey, a distance of about seven kilometres. The journey doesn't take long and Nick usually clocks on at about 5.55.

Nick would like to take the Mersey Ferry to work, but the first ferry doesn't leave the Pier Head until 7.30.

Nick enjoys his job at Albion Engineering because he doesn't do the same work day after day – in fact, he often doesn't know what the day will bring. But there's always plenty to do. Sometimes Nick does routine maintenance work, sometimes he repairs machines and sometimes he helps to install new equipment.

The first thing Nick does when he starts work is to find out if there are any requests for repair work from the factory, stores or offices, either online over the intranet or as hard copy.

2 **Read the text on page 23 and use the numbers 1–8 to put the pictures in the right order.**

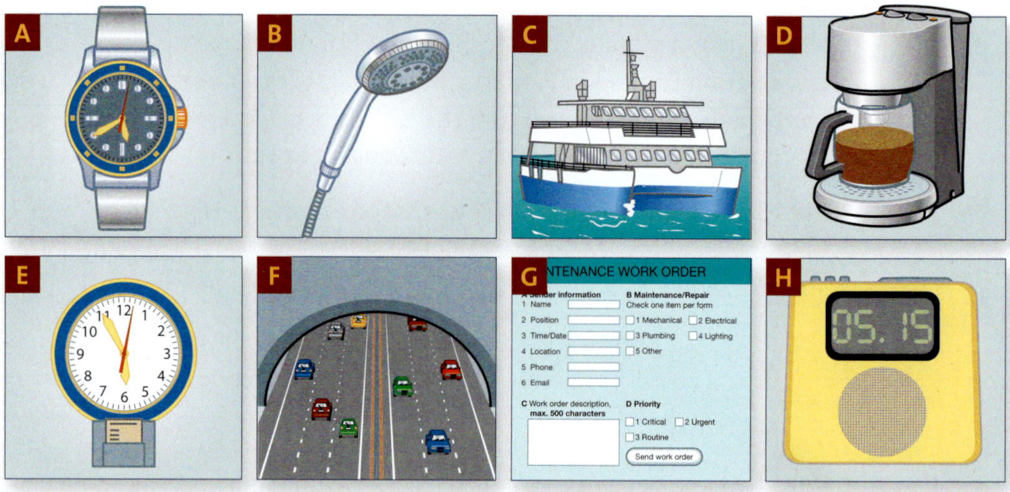

3 **Complete the sentences to make true statements. They are not necessarily in the same order as in the text.**

1. Albion Engineering makes automotive parts for companies ...
2. Nick doesn't find it difficult ...
3. Nick drives to work because the first ferry ...
4. Nick likes the early shift because ...
5. Nick lives in Henry Street in Liverpool because ...
6. Nick's firm works two eight-hour ...

4 **a) Read the text again. Find the English equivalents for the following German words/ expressions. They are in the same order.**

1. Wartungstechniker
2. Frühschicht
3. Schichtzulage
4. Arbeitstag
5. bei Arbeitsantritt stempeln
6. wieder instand setzen
7. einbauen
8. Ausrüstung

b) Now complete the sentences with suitable English words/expressions.

1. You ... to show when you start work.
2. A ... ... repairs and services machines.
3. Do you enjoy your job? Tell me about your ... ...
4. Our firm pays a 12% ... ... to people who do the early shift.
5. Nick sometimes helps to ... new machines.
6. "..." is any tool or machine that helps to do a job.

MAINTENANCE WORK ORDER

A Sender information

1 Name [ ]
2 Position [ ]
3 Time/Date [ ]
4 Location [ ]
5 Phone [ ]
6 Email [ ]

B Maintenance / Repair

Check one item per form.

☐ **1** Mechanical ☐ **2** Electrical
☐ **3** Plumbing ☐ **4** Lighting
☐ **5** Other

C Work order description, **max. 500 characters**

[ ]

D Priority

☐ **1** Critical ☐ **2** Urgent ☐ **3** Routine

Send work order

5 **Sehen Sie sich das Antragsformular an. Benutzen Sie die Kennziffern, um das Formular mit den folgenden Inhalten richtig auszufüllen.**

automatic door, Store 2 • Critical • 803 • Electrical • 17.32, 5 June 20.. • Jack Davis
j.davis@albeng.co.uk • Logistics Manager • Motor cuts out too early •

A.1 *Jack Davis* A.2 ...

6 Tr4 **First, read the word list. Do not just note the meaning of the words, but think about their sound, too. Then look at the ten statements below so that you know what to listen for. Finally, listen to the dialogue and say if the statements are true or false.**

Word list

depend on	*von etwas abhängen*	switched off	*abgeschaltet*
switch	*Schalter*	motor casing	*Motorgehäuse*
complicated	*kompliziert*	damaged	*beschädigt*
urgent	*dringend*	buckled	*verbogen*
spark	*Funke*	faulty relay	*fehlerhaftes Relais*
(to) overheat	*heiß laufen*	repair	*Reparatur*

1. Nick Kelly phones Jack Davis.
2. Nick wants to ask Jack some questions.
3. If the job is complicated, Nick can't do it before about 12.30.
4. Nick has two other critical jobs to do.
5. Nick's first question is about the door.
6. Nick wants to know if the motor overheats.
7. Jack tells Nick that the power is still switched on.
8. Jack says the door goes up and down easily.
9. Nick thinks the problem is just a faulty relay.
10. Nick says repairs like that take a long time.

Das *simple present*

Form

Beispiele	Regeldarstellung
Nick and Jenny **live** in Henry Street. Nick usually **clocks on** at about 5.55.	**Aussagen**: wie der Infinitiv. Ausnahme: Die **3. Person Singular** (*he, she, it*) endet auf **-s**.
I **do not/don't have** a car. Nick **does not/doesn't have** breakfast.	**Verneinung**: Gibt es kein Hilfsverb im Satz, benutzt man *do not/don't* bzw. *does not/doesn't* (3. Person).
Does Jenny **work** at Albion Engineering? **Do** you **work** here in London?	**Fragen**: Gibt es kein Hilfsverb im Satz, benutzt man ebenfalls *do/does*.

Gebrauch

Beispiele	Regeldarstellung
Nick **works** at Albion Engineering. Lessons **begin** at 9 o'clock. A mirror **reflects** the light.	Man benutzt diese Zeit für **dauerhafte Zustände, sich ständig wiederholende Ereignisse bzw. Handlungen** und **allgemein gültige Zusammenhänge**.

■ Das *simple present* steht häufig mit den folgenden Zeitangaben:

– *always, never, often, sometimes, rarely/seldom* (selten)
– *as a rule* (in der Regel), *generally* (im Allgemeinen), *normally, now and then* (hin und wieder), *regularly* (regelmäßig), *usually* (gewöhnlich)
– *every day/week/month/..., every morning/afternoon/...*
– *on Mondays/Tuesdays/...*

1 **Put the verbs into the *simple present*.**

1 Albion Engineering ... ¹ (make) automotive parts. It ... ² (send) its products to car companies all over the world.
2 Nick Kelly ... ¹ (be) a maintenance technician with Albion Engineering and his partner, Jenny, ... ² (have) a job there, too.
3 Nick and Jenny ... ¹ (live) in Liverpool but they both ... ² (work) in Wallasey.
4 Jenny: Nick ... ¹ (get) home at 2.30 but I ... ² (finish) at 5.30. But maybe this ... ³ (be) good. Nick ... ⁴ (do) a lot of the housework before I ... ⁵ (come) home.
5 Manager: Our technicians ... ¹ (be) well qualified and we also ... ² (give) them training.
6 Nick ... ¹ (find) his work interesting because he ... ² (do) different jobs every day. The technicians sometimes ... ³ (do) routine servicing, but they also ... ⁴ (repair) machines and ... ⁵ (install) new ones.

2 Complete the sentences. Look at the examples carefully first.

Examples:

A Peter wants a job with Albion Engineering. (Alan)
B No. Peter *doesn't want* a job with Albion Engineering. Alan *does*.

A The women play computer games at work. (The men)
B No. The women *don't play* computer games at work. The men *do*.

1 A Nick lives in Wallasey. (His parents)
B No. Nick ... in Wallasey. His parents ...
2 A Peter and Alan are engineers. (Jack and Jenny)
B No. Peter and Alan ... engineers. Jack and Jenny ...
3 A Sally comes from Wales. (Megan)
B No. Sally ... from Wales. Megan ...
4 A The computers are made in Germany. (The copiers)
B No. The computers ... made in Germany. The copiers ...
5 A Albion Engineering wants to go to Poland. (Mersey Metals)
B No. Albion Engineering ... to go to Poland. Mersey Metals ...
6 A Jack watches too much TV. (His children)
B No. Jack ... too much TV. His children ...
7 A Nick enjoys cycling in the summer. (His girlfriend Jenny)
B No. Nick ... cycling in the summer. His girlfriend Jenny ...
8 A Susan speaks German. (Ann)
B No. Susan ... German. Ann ...

3 Use the sentence elements to ask questions. Add any missing words. Look at the example carefully first.

Examples:

– Nick + work/Wallasey? *Does Nick work* in Wallasey?
– Jenny + be/Nick's girlfriend? *Is Jenny* Nick's girlfriend?

1 Ann + speak/German/well?
2 Why/be + Jack/Cologne?
3 Nick and Jenny + work/Albion Engineering?
4 When/Nick + start/work?
5 How much/these laptops + cost?
6 Where/Halina Piatkowski + come from?
7 Halina and Aldona + be/Polish?
8 Bob and Ann + have/family?
9 How/you + install/new software?
10 How often/Jack + do/overtime?

This conversation between Alan Townsend, production manager at Albion Engineering, and Greg Hill, the firm's safety officer, is about a bad accident in the factory. Frank Richter, an installation technician from Germany, was badly hurt when he fell off a scaffold tower. The question now is: Who or what caused the accident? Did Greg Hill make mistakes? Or was Frank Richter at fault?

mobile scaffold tower

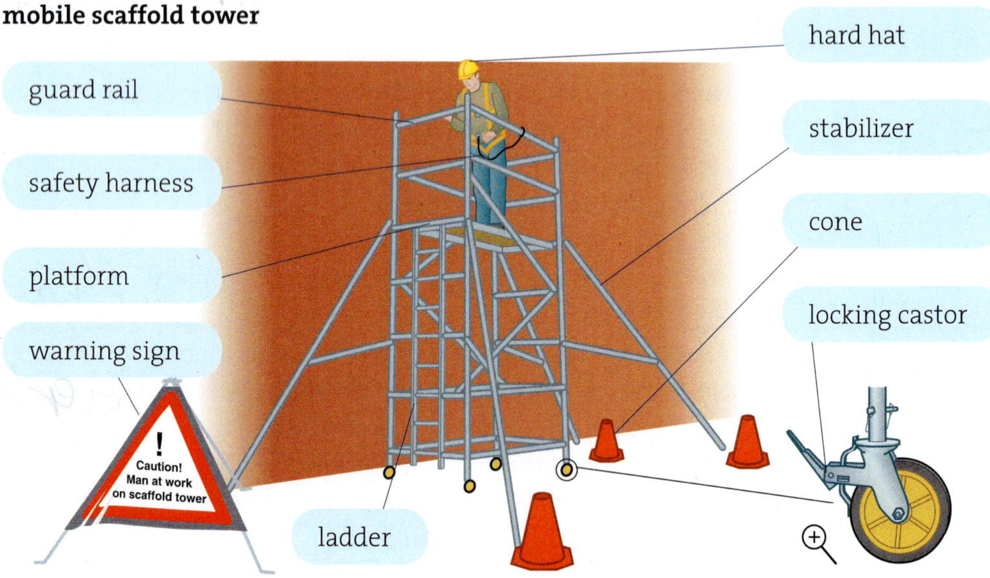

1 **Link the sentence parts 1–8 with the sentence parts a–h to make true statements about the mobile scaffold tower and the surrounding area.**

1 The tower supports a platform so that
2 Workers get on to the platform by using
3 There are guard rails round the platform
4 A safety harness fixed to the guard rail
5 Long stabilizers at each corner stop
6 The tower is easy to move
7 Red cones mark the area around the tower
8 Warning signs around the tower also

a a ladder fixed to the side of the tower.
b catches the worker if he falls.
c on its locking castors at each corner.
d people can work safely at a height.
e tell other people to be careful.
f the tower from swaying (*schwingen*).
g to stop people from coming too close.
h to stop workers from falling off it.

1 *d*
2 ...

castors	floor	hard hat	safe	shift
cones	forklift	hit	safety	shop
fell	guard rail	metres	safety harness	warning signs

2 Tr 5 **Work with a partner. Listen to the dialogue and complete it with words or expressions from the box.**

forklift (truck)

Alan Okay, Greg. It's about the accident with a scaffold tower in the pre-assembly shop on 21 September. What happened exactly?

Greg Yeah, well, it was about 2.10 p.m., just after the late ... [1] started. A ... [2] came ...

Alan Who was the driver?

Greg Ali. Ali Hussan. He's a good man. He's been with us for over ten years now.

Alan Okay. Go on.

Greg Well, Ali drove through the swing doors from the parts store into the pre-assembly ... [3] and hit the scaffold tower. Frank fell nine ... [4]. He was badly hurt. He may never walk again.

swing doors

Alan Yes, but *how* did it happen, Greg? I'm sorry about the guy's injuries, of course, but this could cost the firm millions. Christ, the guy may never work again.

Greg Yes, but we are insured, aren't we, Alan?

Alan Sure, but only if the accident wasn't our fault. We guaranteed a "... [5] working environment", don't forget. That's your job.

Greg Yes, it is – and I did it. There was nothing wrong with our ... [6] measures, I promise you that. There were two big ... [7] in the parts store and ... [8] round the tower. And Ali and the other drivers knew all about it as well. We didn't miss a thing, Alan.

Alan Okay, I believe you. But the fact is that ... what's his name again?

Greg Frank. Frank Richter.

Alan Yes, well, this Frank fell nine metres on to a concrete ... [9]. How was that possible? Did he have a safety harness?

Greg Well, we gave him a safety harness, but ...

Alan So ... no ... [10], right? What else?

Greg Look, Alan, I don't want to ... Well, I mean the poor guy's ...

Alan Get on with it, Greg. For God's sake ...

Greg Well, Frank sometimes took short cuts, Alan. Safety costs time and trouble, that's the problem. In this case, well, Frank didn't have a safety harness or a ... [11], he took off the top ... [12] and the tower castors weren't locked.

Alan I see. So when the forklift truck ... [13] the scaffold tower, the tower suddenly moved and Frank ... [14], right?

Greg Yes, that's about it, Alan.

Alan Okay, but why did the forklift hit the tower? How did that happen?

Greg Well, that really was just bad luck – an accident. Ali thought somebody was about to walk in front of his forklift. He swerved to the right and hit the tower. He didn't hit it hard – just kissed it.

Alan I see. Now for the big question Greg. If Frank had locked the ... [15], what would have happened then?

Greg Nothing. No accident, no poor Frank.

Alan Okay then, Greg. That doesn't sound too bad. Can you write an accident report for our insurers and for Frank's firm, please?

Greg Okay. I'll get on with it at once.

3 **Say if the statements are true or false. Correct the false statements.**

1 The forklift came from the parts store into the pre-assembly shop.
2 The forklift hit the scaffold tower and Frank fell six metres.
3 Frank was not badly hurt in the accident.
4 Alan says the accident could cost the firm millions.
5 Greg is the firm's quality manager.
6 Frank didn't have a safety harness on at the time of the accident.
7 Frank sometimes took short cuts because safety costs time and trouble.
8 The driver swerved to the left and hit the tower.

4 **Complete the compound with the missing words from pages 28–30.**

1 accident ...
2 pre-assembly ...
3 ... truck
4 ... rail
5 parts ...
6 production ...
7 ... harness
8 safety ...
9 scaffold ...
10 ... cut
11 swing ...
12 ... environment

5 Read the description of how the accident happened again. Draw a labelled sketch.

6 There are ten factual mistakes in this extract from Greg's report. Correct them.

Accident Report

Place	final assembly shop [1] just inside door from pre-assembly [2]
Time / Date	14.10, 23 [3] September 20...
People involved	Jack Cooper [4], forklift truck driver Frank Richter, quality inspector [5], MAG-Fördertechnik, Essen, Germany
Safety measures	Warning sign on parts store side of swing doors Scaffold tower with stabilizers Scaffold tower site marked with flashing yellow lights [6]
	hard hat safety net [7] guard rails locking jacks [8]
Description	Ali Hussan drove a forklift truck through the rolling door [9] between the parts store and the pre-assembly shop. The forklift truck hit a mobile platform ladder [10] located approx 8 metres from the doors. This happened when Mr Hassan tried to avoid somebody who he thought was about

7 First, look at the grid. Then listen to the extract from a radio talk and put the missing figures into the grid at letters a–d.

Tr 6

Accidents in the home	
Total accidents	approx 160,000
Total falls	approx – a –
– down stairs/steps – out of bed – from chairs/sofas	approx – b – approx – c – approx – d –
– other falls	approx 8,500

Das *simple past*

Form

Beispiele	Regeldarstellung
The tower suddenly **moved**. Ali **drove** through the swing doors.	**regelmäßige Verben**: Anhängen von -**ed** an den Infinitiv. **unregelmäßige Verben**: Sonderform, die man sich merken muss, s. Unregelmäßige Verben, S. 162.
We **did not/didn't miss** a thing. Frank **did not/didn't have** a hard hat.	**Verneinung**: Man benutzt **did not/didn't + Infinitiv** mit allen Personen.
Did Frank **have** a safety harness?	**Fragen**: Gibt es kein Hilfsverb im Satz, benutzt man ebenfalls **did**.
Who hit the scaffold tower? **What happened** exactly?	! Ist **who** oder **what** das **Subjekt**, bildet man die Frage **ohne** Hilfsverb.

Gebrauch

Beispiele	Regeldarstellung
When the forklift truck **hit** the scaffold tower yesterday, the tower **moved** and Frank **fell** nine metres.	Man benutzt diese Zeitform, um **völlig abgeschlossene** Situationen oder Handlungen in der Vergangenheit zu schildern.

■ Das simple past steht häufig mit den folgenden Zeitangaben:

- *yesterday, the day before yesterday* (vorgestern)
- ***last** night/month/summer/July/Easter/...*
- *two/three/... minutes/hours/weeks/years **ago***
- *in 2001/...*
- *in those days/..., when she was a girl, when he worked in New York, ...*

1 Work in pairs. Look at Nick's worksheet and tell your partner what Nick did last Wednesday. Swap roles at least once. Use the *simple past*.

Albion Engineering Limited – Maintenance Department job sheet

Name: *Nick Kelly* Date: *10 May 20..*

Please fill in jobs etc. to nearest half hour. This is not a timesheet.

6.00	*check requests for maintenance work*	*9.00*	*repair swing door, pre-assembly shop*
6.30	*go to Greg Hill's office to repair window*	*9.30*	*install new motor, parts store conveyor*
		10.00	*fetch forklift truck for servicing*
7.00	*change light bulb, canteen*	*11.00*	*inspect all fire extinguishers*
7.30	*clean filters, extractor fans ()*	*12.00*	*write memo about servicing office machines*
8.00	*...*		
8.30	*...*	*12.30*	*phone Carr & Co. about spare parts*

1 *Nick checked requests for maintenance work at 6.00.*
2 *At 6.30 he ...*

2 Complete the sentences. Use the short form of the *simple past*.

1 Unfortunately, Frank *didn't wear* a safety harness or hard hat.
2 He ... (not lock) the castors because he wanted to save time.
3 He ... (not want) to climb down all the time to move the tower.
4 Ali's forklift truck ... (not hit) the tower very hard.
5 Greg and his men ... (not forget) any safety measures.
6 Alan and Greg ... (not think) the accident was their fault.
7 Alan: Our insurers ... (not pay) Frank any money because he ... (not obey) the safety rules.
8 Ali ... (not drive) too fast. He's always very careful.
9 Greg: We ... (not give) Frank safety equipment just for fun, you know.
10 Greg ... (not write) the report. His assistant did.

3 Your partner is Ali. Use the German prompts to ask and answer questions about the accident. Swap roles at least once. Use the *simple past*.

Sie möchten wissen ...

1 ... wer für die Arbeitssicherheit die Verantwortung trug (*be in charge of*).
A *Who was in charge of job safety?*
B *Greg Hill.*
2 ... wann und wo der Unfall passiert ist.
3 ... wie viele Männer auf dem Turm waren.
4 ... warum Ali nach rechts ausgeschwenkt ist.
5 ... wie tief (*far*) Frank gefallen ist.
6 ... welche Sicherheitsmaßnahmen es gab.
7 ... ob Frank eine Sicherheitsausrüstung trug.
8 ... wer für den Unfall verantwortlich war.

Asking the way

a satnav

This is part of the street plan of a town somewhere in a German-speaking country. The van you can see at bottom right is driven by Tom Bates from Britain. Tom is on his way to make a delivery to a company in the town called MAG-Fördertechnik, but he has a problem. His satnav isn't working, and he doesn't have a street plan, either. He is hopelessly lost in a strange town, but luckily there's an easy answer to his problem. Tom uses his mobile to phone MAG to ask for directions about how to reach them.

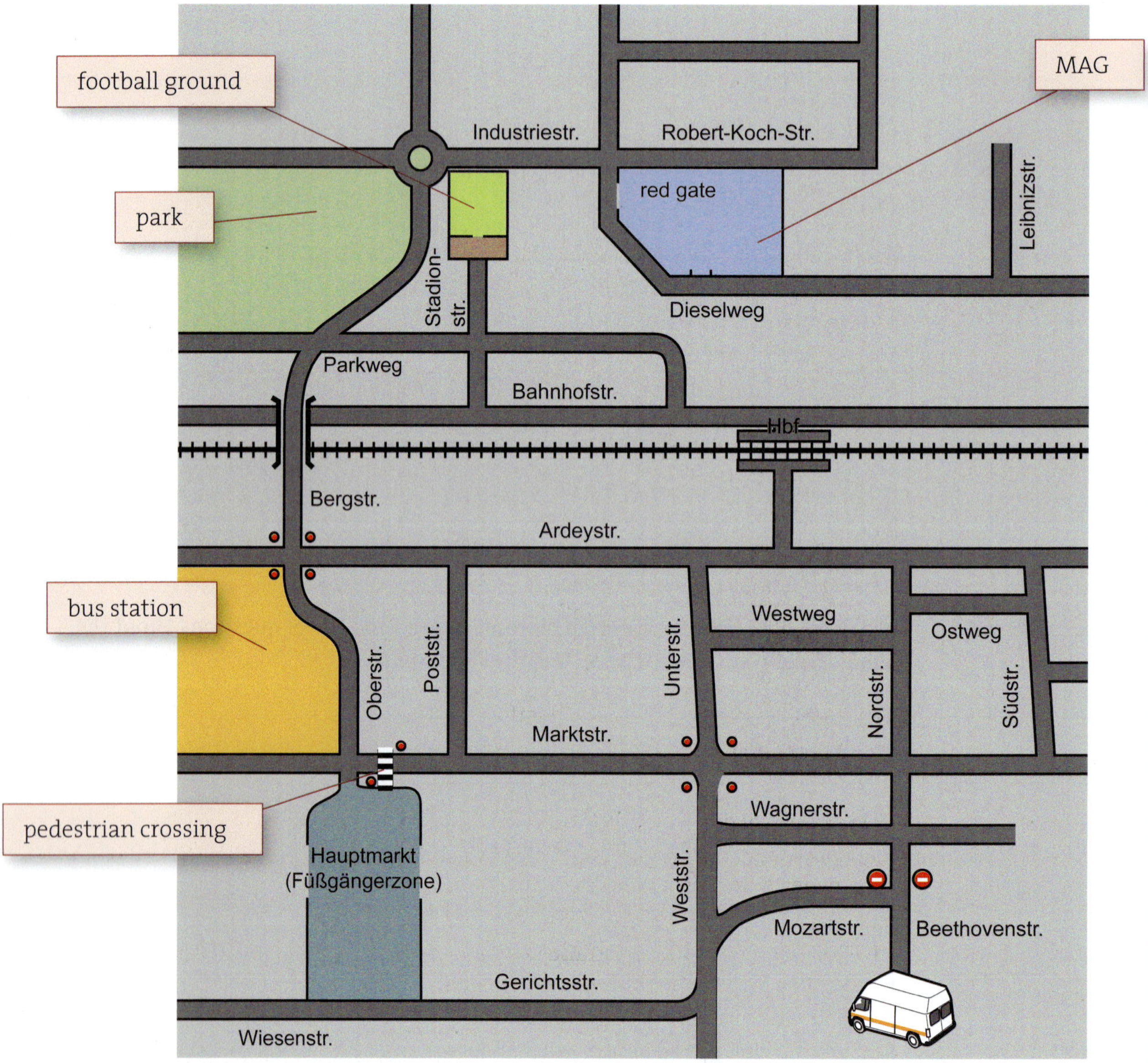

1 **Work with a partner. Listen to or read the dialogue and put the eight instructions in the correct order. Look at the street plan on page 34.**

Tr 7

A Drive up Marktstraße.
B Go straight on into Bergstraße.
C Please use the red gate.
D Take the first right again into Dieselweg.
E Turn left there into Marktstraße.
F Turn right at the roundabout.
G You must turn immediately right into Oberstraße.
H You must turn left into Mozartstraße.

1 *H*
2 ...

Gabi Guten Tag. MAG-Fördertechnik. Mein Name ist Gabi Taubitz. Was kann ich für Sie tun?
Tom Oh ... Um ... Look, I'm sorry but may I speak English, please?
Gabi Sure. What can I do for you?
Tom Well, my name's Tom Bates. I'm on my way to you with a delivery, but my satnav is down and I don't have a street plan. I can't find the way.
Gabi No problem. Where are you now?
Tom Um ... I'm in Beethovenstraße.
Gabi Oh, dear. Can you tell me what you can see in front of you, please?
Tom Well, I can see a No Entry just down the road.
Gabi Okay. Well, this may be a little complicated, I'm afraid. Do you have something to write with?
Tom Sure. Go ahead.
Gabi Well, go to the No Entry sign and there you must turn left into Mozartstraße. It's the only way you can go.
Tom Okay. Left into Mozartstraße.
Gabi Well, at the end of Mozartstraße there's a T-junction. Turn right there into Weststraße. You mustn't turn left there or you could get into a one-way system.
Tom Okay. I have that. Don't turn left at the T-junction.
Gabi Fine. Now, after about 200 metres you come to some traffic lights. Turn left there into Marktstraße.
Tom Left into Marktstraße.
Gabi Good. Now, drive up Marktstraße and after about 300 metres you come to a big pedestrian crossing with traffic lights, okay?
Tom No, not really. Can you repeat that, please?
Gabi Yes, well, go up Marktstraße until you come to a big pedestrian crossing with lights. There's a bus station over to your right.
Tom Okay. Thanks. I've got that now.
Gabi Fine. Now, drive over the pedestrian crossing and then turn immediately right

into Oberstraße. Be careful there. Do not drive straight on there or you'll have awful problems.

Tom Got that. Right into Oberstraße.

Gabi Now drive up Oberstraße to the traffic lights. Go straight on there into Bergstraße.

Tom Straight on into Bergstraße. I have that.

Gabi Fine. Well, now you have to go along Bergstraße for about a kilometre until you come to a roundabout. There should be a football ground on your right and a park on your left.

Tom Roundabout, football ground and park, right?

Gabi That's right. You're nearly there now. Turn right at the roundabout and after about 150 metres take the first right again into Dieselweg. MAG's the first on the left. You really can't miss it.

Tom Great. Thanks a lot for your help.

Gabi Don't mention it. You're welcome. Oh, and please use the red gate. You can't reach the stores through the green one, you see.

Tom Fine. I'll do that. Bye now.

Gabi Goodbye, Mr Bates – and good luck.

2 **Find the English equivalents of the German sentences in the dialogue.**

1 Könnte ich vielleicht Englisch sprechen?
2 Was kann ich für Sie tun?
3 Kein Problem.
4 Dies könnte ein bisschen kompliziert werden.
5 Fangen Sie an.
6 Das habe ich verstanden.
7 Können Sie das bitte wiederholen?
8 Vielen Dank für Ihre Hilfe.
9 Keine Ursache.
10 Bitte schön.

3 **Read the two dialogues with a partner. Then use the street plan on page 37 to ask the way and give directions. You are at the *town hall*.**

A Excuse me, please. Can you tell me the way to the technical college?
B Sure. Just walk down Essex Road and take the first right into Windsor Road. Go down Windsor Road and the technical college is on the corner of Windsor Road and Church Street. It's opposite the police station. You can't miss it.
A Fine. Thanks for your help.
B You're welcome.

A Excuse me. Is there a youth hostel in the town?
B No problem. Are you on foot?
A Yes, I am.
B Fine. Then walk along Albert Road and straight on over the roundabout into Brick Lane. You'll see the youth hostel on the left after about 200 metres. It's on the corner of Brick Lane and Station Road. It isn't far from here.
A Thanks a lot. Goodbye.

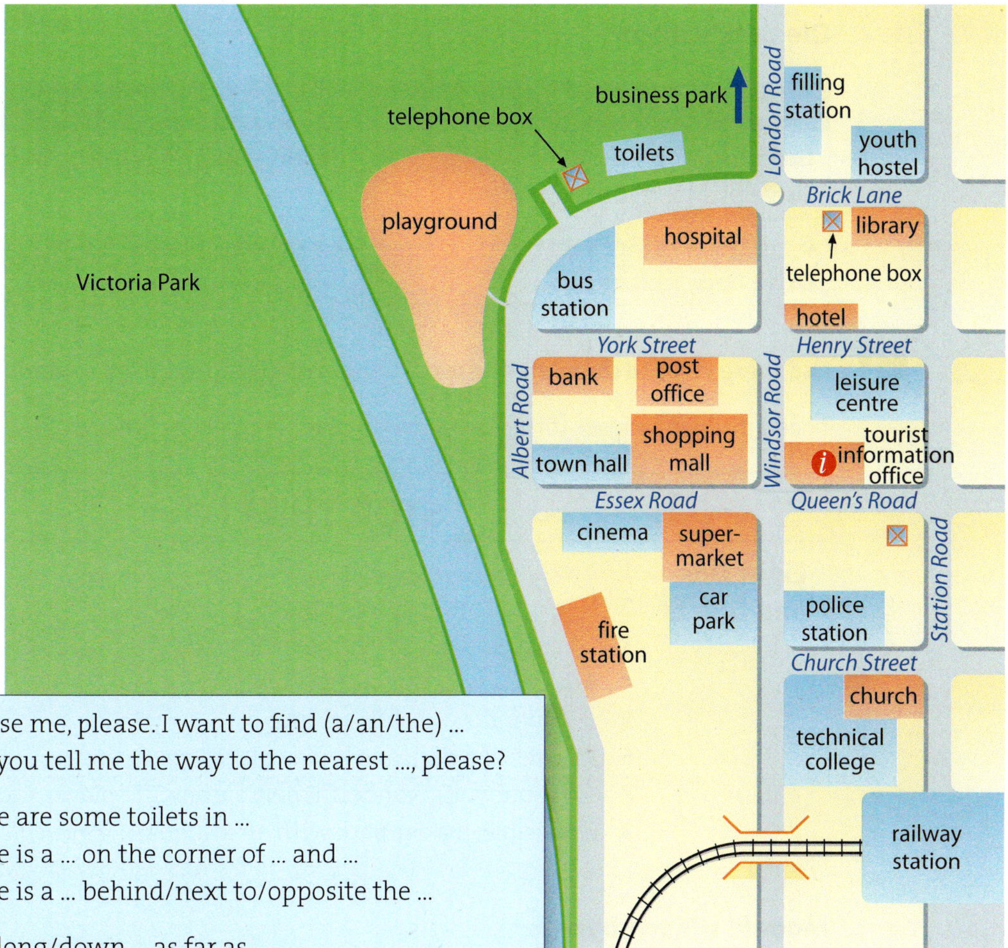

Excuse me, please. I want to find (a/an/the) ...
Can you tell me the way to the nearest ..., please?

There are some toilets in ...
There is a ... on the corner of ... and ...
There is a ... behind/next to/opposite the ...

Go along/down ... as far as ...
Turn left/right into ...
Take the first/second/... road on the left/right.
Go straight on past ... as far as ...
The ... is on your left/right.

4 **Tell your partner how to reach various places. You are at the tourist information office.**

5 **Read the sentences 1–6. Then listen to the dialogue and fill in the missing details.**

Tr 8

1. Jenny wants to go to the new ... in St Helens.
2. The multiplex has ... screens, ... bars and ... restaurants.
3. Jenny doesn't want to go to St Helens by ...
4. St Helens is ... stops from Preston Road.
5. It's a ... minute walk to the supertram stop in Preston Road.
6. In the end, Nick and Jenny take a ... to Preston Road.

Die Befehlsform

Beispiele	Regeldarstellung
Go to the No Entry sign. **Drive** up Marktstraße.	**Bejahte** Aufforderungen: Infinitiv (ohne *to*)
Go to the No Entry sign. **Drive** up Marktstraße.	**Verneinte** Aufforderungen: *Do not/Don't* + Infinitiv (ohne *to*)

1 **Use these verbs to complete the leaflet. Use negative forms in gaps marked *.**

boast (*prahlen*) · close · forget (*vergessen*) · give · leave · lock (*abschließen*) · park · take · use

STOP, THIEF!

Eight Dos and Don'ts

1 Always ... your vehicle.
2 Never ... your keys to strangers (*Fremde*).
3 ... * cameras etc. in your vehicle.
4 Always ... your documents with you.
5 ... all windows. ... * the sunroof!
6 Thieves like dark and lonely places, so ... * there.
7 ... * about your wonderful new satnav, CD-player, etc.
8 If possible, ... a car park with an attendant (*bewachter Parkplatz*).

Modale Hilfsverben

Beispiele	Regeldarstellung
can/could	*können/könnten*
I **can see** a No Entry sign. You **could get** into a one-way system.	Man benutzt *can/could* im Englischen genau wie „können/könnten" im Deutschen.
have to/must	*müssen*
You **have to go** along Bergstraße. You **must turn** left into Mozartstraße.	In **normalen Situationen** benutzt man *have to*. Möchte man aber zeigen, dass es sich um eine **zwingende Angelegenheit** handelt, verwendet man *must*.
mustn't	*nicht dürfen*
You **mustn't turn** left there.	Mit *mustn't* drückt man ein **Verbot** aus.
may	*dürfen*
May I speak English? I **may be** late tomorrow, Nick.	Man benutzt *may*, um Erlaubnis (dürfen) und Möglichkeit (vielleicht) auszudrucken.

2 **Look at the signs. Use *must*, *can*, *mustn't*.**

1 You ... wear a hard hat in this area.

2 People ... smoke in this building.

3 You ... be careful here.

4 We ... drive as fast as we like now.

5 You ... park here.

6 We ... park here but we ... buy a ticket.

3 **Translate the highlighted parts of the sentences into English.**

1 Könnte ich vielleicht Englisch sprechen?
2 Links können Sie das Rathaus sehen.
3 An der Kreuzung müssen Sie geradeaus fahren.
4 Pass auf! Du könntest sonst Probleme bekommen.
5 Er darf nicht so viel trinken.
6 Heute muss ich mit dem Bus fahren.
7 Dies könnte etwas kompliziert werden.
8 Hier dürfen wir nicht parken.

In this unit, you are going on a tour of a manufacturing company, Albion Engineering Ltd. During such tours, you generally visit all the important departments of the company in logical order. In practice, this means that you follow the progress of material from goods inward through assembly and final inspection to packaging and dispatch. After looking at the manufacturing process, you may go on to visit non-manufacturing departments like the drawing office and administration.

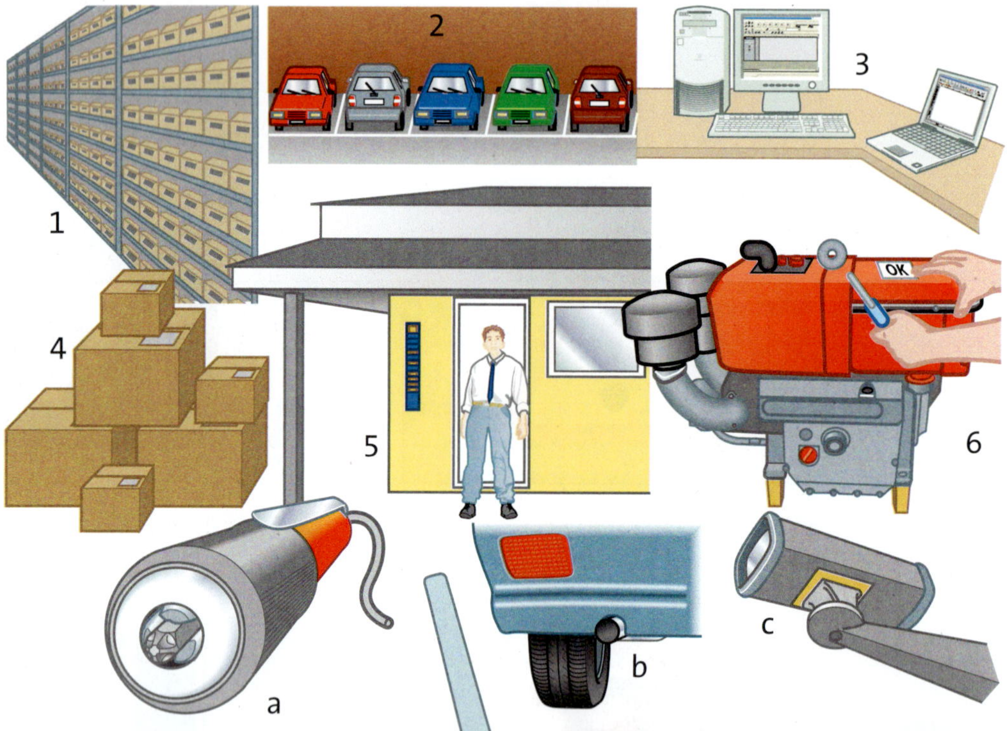

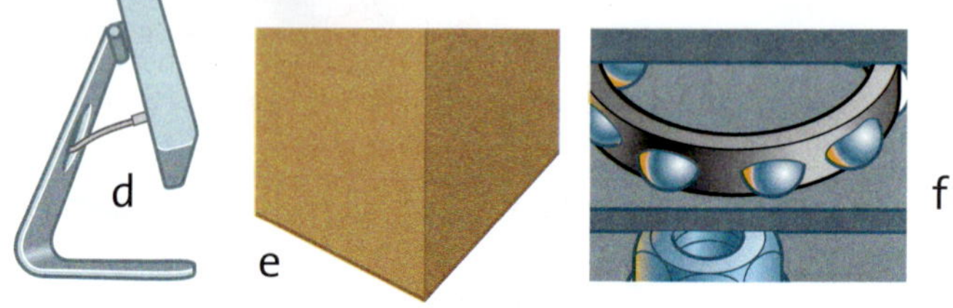

car park, dispatch, drawing office, final assembly, parts store, plant security office

1 **First, link the places (1–6) with the correct caption (*Bildunterschrift*) from the blue box. Then link the places with the objects (a–f) that go with them. The words in the pink box will help you to do this, but be careful as some don't fit at all.**

The first/... picture is of a parts store/... I think this picture goes with picture *f*/..., which shows/seems to show components/parts on shelves/...

cardboard box
exhaust outlet
monitor
screwdriver
shelving
stabilizer
suitcase
surveillance camera
TV screen
water tap

2 **Look at the plan of the Albion Engineering plant. Then answer the questions.**

1 Where did Frank Richter's accident (Unit 4) happen?
2 Where does Nick Kelly park?
3 Which other departments must Nick pass to reach the Maintenance Department?
4 Where does final inspection take place (*stattfinden*)?
5 You want to deliver (*anliefern*) some parts. Where do you go to?
6 You see a suspicious stranger (*verdächtiger Fremder*) near R&D. Where can you get help?
7 Where do you go to get spare parts?
8 Where do the trainees work at Albion Engineering?

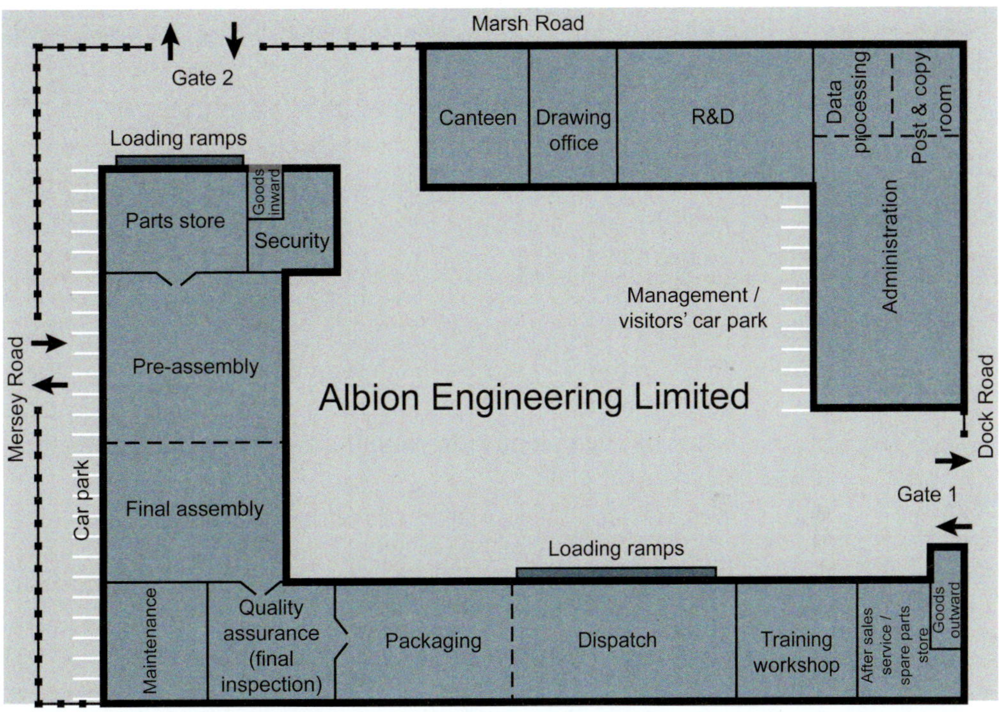

administration	*Verwaltung*	management car park	*Parkplatz der Geschäftsleitung*
after-sales service	*Kundendienst*	packaging	*Verpackung*
canteen	*Kantine*	parts store	*Bauteillager*
car park	*Parkplatz*	post and copy room	*Post- und Kopierstelle*
data processing	*EDV*	pre-assembly	*Vormontage*
dispatch	*Versand*	quality assurance	*Qualitätssicherung*
drawing office	*Zeichenbüro*	R&D	(**R**esearch **&** **D**evelopment) *Forschung und Entwicklung*
final assembly	*Endmontage*		
final inspection	*Endkontrolle*		
gate	*Tor*		
goods inward	*Warenannahme*		
goods outward	*Warenausgabe*	security	*Werkschutz*
loading ramp	*Laderampe*	spare parts store	*Ersatzteillager*
maintenance	*Instandhaltung*	training workshop	*Lehrwerkstatt*
		visitors' car park	*Besucherparkplatz*

Tr9 A tour for apprentices

Bill Atkins is training manager at Albion Engineering. He is showing eight young people round the plant. They all hope to get an apprenticeship, but there are only four places. Then Bill has a difficult job. He has to decide who gets an apprenticeship, and who will be disappointed. One important point is how much interest applicants show during a tour of the plant.

Bill Okay then. Listen carefully, please. We are now in the middle of the yard. It may look very empty now, but trucks are so big nowadays that they need plenty of space. Now, let's look towards Mersey Road. The long building on the left and the shorter one in front of us make up the assembly plant and logistics. Now, turn round and look towards Dock Road, please. ... Are you with us, Peter? I said Dock Road, not Marsh Road.

Peter Oh, sorry. I don't come from here. It's all new to me.

Bill No problem. Anyhow, this L-shaped building is the main office building. The shorter building in front of us is administration and the longer one is made up of R&D, ...

Peter Excuse me ...

Bill Yes, Peter?

Peter Well, what's R&D?

Bill R&D stands for "research and development". This is the first step in producing a new or improved product. From there, a new product goes to the drawing office, through production planning and, finally, to the assembly plant, okay?

Peter Okay. Thanks.

Bill Well, you won't have much to do with the office building – except for the canteen, of course. Now, are there any more questions so far? No? Okay, then. If there are no more questions, let's walk slowly across the yard to Gate 2. Oh, and please keep to the walkways, all right? That's the very first safety rule. Never leave the walkways. If you do, there could be a ...

... Now, here we are at the loading ramps of goods inward. All parts and components from our over 120 suppliers come into the plant here. Everything is thoroughly inspected, sorted and then stored in the parts store until ...

Amar Excuse me. I have a question.

Bill Fine. Go ahead. Um ... Amar, is it?

Amar That's right. Amar Gupta.

Bill What's your question, Amar?

Amar Well, I just wondered ... I mean, like, do you do any manufacturing at all here?

Bill That's a good question. The answer is that we stopped manufacturing here in Wallasey five years ago. It was too expensive, I'm afraid.

Amar What, so everything's bought in, right?

Bill That's right. We are an assembly plant. It's tough, I know, but to stay in business, we must be competitive in a global marketplace.

3 **Link what Bill says (1–8) with a place from the box. Note that three places do not fit at all.**

administration after sales service canteen dispatch drawing office final inspection goods inward goods outward management car park parts store security

1. All finished products must go through here before going into packaging.
2. All trucks collect goods here.
3. Data processing is in that building over there.
4. Components go from here through the double doors into pre-assembly.
5. You can see all parts of the plant on video monitors in here.
6. You can get a drink here if you like.
7. This department has the wrong name, really. It's all digital nowadays, of course.
8. The bosses get very angry if people park here – so don't!

4 Tr 10 **First read the word lists. Then listen to what the people say about their jobs. Finally, use words from the box to say which department they work in. Be careful. Four words don't fit at all.**

1. skilled (job) *qualifizierte (Arbeit)*; cook *kochen*; serve *servieren*; washing-up *Spülen*
2. CAD **c**omputer-**a**ided **d**esign; suppose *annehmen*; efficient *leistungsfähig*
3. surprised *überrascht*; search *durchsuchen*; less threatening *weniger bedrohlich*; get results *Erfolg haben*; simply *einfach*; willing *bereit*
4. reject *ablehnen*; worse *Schlimmeres*; properly *richtig*; close *schließen*

canteen data processing drawing office maintenance packaging parts store quality assurance security

5 ***Rollenspiel***

Sie sind Bill Atkins.
Benutzen Sie das Material, um den Rundgang fortzusetzen.
Besuchen Sie mindestens zwei Abteilungen.

This is the quality assurance department/... That is maintenance/... over there. We're now standing in the ... After the ..., products go into the ... Here you can see ... On our right/left ... To the right/left of ... In front of us ... Between ... and ... there is the ... The ... is next to there ... Over there, you can see ...

Adjektive und Adverbien

Beispiele	Regeldarstellung
Nick lives in an **old part** of town. The **early shift** is from 6 a.m. to 2 p.m. That doesn't **sound bad**.	**Adjektive** beschreiben **a**) Nomen und **b**) einige wenige Verben wie *look* (aussehen), *seem* (scheinen), *smell* (duften, riechen), *sound* (klingen) und *taste* (schmecken).
Let's **walk slowly** to Gate 2. Frank was **badly hurt** in the accident.	**Adverbien** beschreiben **a**) Verben und **b**) Adjektive. Die meisten Adverbien enden auf **-ly**.
The canteen is in the **longer** building. Nick gets home **earlier** than Jenny. Wallasey is **more expensive** than Liverpool. Tom must drive **more carefully**. Ali is **good**. He's the **best** driver we have.	**einsilbige** Adjektive und **zweisilbige** Adjektive auf **-ly**: Steigerung mit **-er**/**-est**. **mehrsilbige** Adjektive und **Adverbien auf -ly**: Steigerung mit **more**/**most**. **unregelmäßige** Steigerung: good – better – best; bad – worse – worst.

1 **Choose the correct word, a) adjective or b) adverb.**

Instructions for use

1 Unpack your a) new – b) newly [1] **STAR 2000 DVD-Recorder** a) careful – b) carefully [2]. Do not use a knife!

2 Check the contents of the carton a) thorough – b) thoroughly [3] (see packing list). If it is a) incomplete – b) incompletely [4], please contact your dealer a) immediate – b) immediately [5].

3 Place the recorder a) careful – b) carefully [6] on a a) flat – b) flatly [7] and a) firm – b) firmly [8] surface. Do not place in a) direct – b) directly [9] sunlight.

4 Connect the recorder to a TV with a SCART jack. Be a) sure – b) surely [10] to connect the cable a) correct – b) correctly [11] (see drawing). The cable will connect a) easy – b) easily [12] – never use force!

5 Connect the power cable first to the recorder and then to the power point. This order is a) important – b) importantly [13] to protect a) electronic – b) electronically [14] components.

6 Turn on the recorder either by pressing the a) round – b) roundly [15] ON/OFF button or with the a) remote – b) remotely [16] control handset.

2 **Put the words on the left into the sentences. Use the adverb form where necessary.**

slow	**1**	The apprentices walked ... to Gate 2.
quick, careful	**2**	Ali didn't drive ... He's always very ...
exact	**3**	Alan: Greg, please tell me ... what happened.
careful	**4**	Bill told the boys to listen ...
bad	**5**	Frank was ... hurt when he fell from the tower.

Fragen

Entscheidungsfragen und Kurzantworten	
– **Is Bill Atkins** in his office?	– **No**, he **isn't**.
– **Can I speak** to Mr Townsend now?	– **Yes**, you **can**.
– **Does Bill** always **instruct** the apprentices?	– **No**, he **doesn't**.
– **Did Amar get** a job?	– **Yes**, he **did**.

3 **Form questions with *do/does, did*.**

Example: – Nick Kelly + live in Henry Street?
Does Nick Kelly live in Henry Street?

1. you + see the football match last Saturday?
2. the apprentices + go to school every Thursday?
3. I'm sorry, Mr Atkins. I + understand you correctly yesterday?
4. Amar and Peter + work for Albion Engineering now?
5. Jenny + have lunch in the canteen every day?
6. the firm + stop manufacturing five years ago?

Fragen mit einem Fragewort	
– **Where can I find** a screwdriver?	– **How do you switch off** this machine?
– **Why were you** late yesterday, Peter?	– **When did Amar start** work at Albion?

4 **Form questions by asking about the highlighted part of the sentences.**

Example: – Nick works in Wallasey.
Where does Nick work?

where?	**1**	Albion Engineering's car park is in Mersey Road.
who?	**2**	Bill Atkins is Albion's training manager.
how many?	**3**	Albion has 12 apprentices at the moment.
how long?	**4**	The course takes three years.
how often?	**5**	The apprentices go to school on one day a week.
why?	**6**	Two apprentices went home early yesterday because they were ill.
how?	**7**	Frank was hurt when he fell from a scaffolding tower.
when?	**8**	The accident happened at about 2 o'clock.

Warm-up

Björn Drexler is planning to work in an English-speaking country for a few months. He wants to improve his English and get some international experience. He thinks this will give him more career chances in the 'global village'.

Björn reads the job ads in the newspaper and he often visits online job sites. One day he finds an interesting ad at an Irish site, www.jobs.ie. Celtic Security Systems Limited is looking for installation technicians to work all over the EU. The job starts with three months' training at the company's headquarters in Dublin.

Link the photos or realia 1–8 to the captions A–H in the box.

A government job centre
B job offer card
C newspaper
D company website
E school careers advice
F private employment agency
G tip from friend
H internet jobs site

1

2

3

4

5

6

7

8

An internet job ad

Jobseekers | **Employers**

Home | Search for jobs | Jobs by email | Build your CV | Upload your CV | Track your applications

Celtic Security Systems Limited

Dublin, Ireland
Website: www.css.ie

Installation technicians

Celtic Security Systems Limited is Ireland's most successful manufacturer and supplier of security systems, from automatic gates to advanced electronic surveillance devices.

Owing to a big growth in our export business, we are now looking for **installation technicians** to install and service our equipment throughout the EU.

You will join a company which is the market leader in Ireland and which is now rapidly expanding into the European market. CSS employs 580 people at its head office and manufacturing plant in Dublin, and has associates in 24 countries.

The position of installation technician offers excellent rewards, made up of a generous salary, paid overtime and an annual performance bonus. Three weeks' vacation, free health insurance and a free pension plan are further attractions.

Successful applicants will receive three months' training in Dublin before they begin work in their home countries.

You are a qualified electrician or electronics technician and have at least three years' industrial experience.

You have some English and you can also speak at least one other major European language.

While you will be part of a successful team, you can also work alone and on your own initiative.

The position in brief:

Location:	Dublin, then home country in EU
Payment:	Generous salary; paid overtime; performance bonus
Perks:	Free health insurance, free pension plan; current model VW Passat
Category:	Manufacturing / Engineering
Posted:	16.03.20...
Contact:	Email: tara.ryan@hr.css.ie
Post:	Ms Tara Ryan, Human Resources, Celtic Security Systems Limited Unit 12–16, Jamestown Business Park Dublin 11, Ireland
Phone:	+353-1-8654024-12

Mr Kelly	['mɪstə]	Herr Watson	männlich, verheiratet oder ledig
Ms Ryan	[mɪz, məz]	Frau Ryan	weiblich, verheiratet oder ledig
Mrs Evans	['mɪsɪz]	Frau Evans	weiblich, verheiratet

■ Besonders im Berufsleben wird die neutrale Form **Ms** für Frauen benutzt.

1 **a) Look at the ad on page 47. Find the English equivalents for these German words. They are in the same order.**

1 *Monteure/Monteurinnen* **2** *Sicherheitssysteme* **3** *Lieferant*
4 *Überwachungsgeräte* **5** *Anlagen* **6** *Bewerber/Bewerberinnen* **7** *Schulung*
8 *Elektroniker/in* **9** *Berufserfahrung in der Industrie* **10** *selbstständig*

b) Complete the sentences with words or expressions from a).

1. Don't ask me. You should learn to work ... your ... ...
2. Manufacturers of ... ... are doing well because of terrorism and crime.
3. You see ... ... like video cameras and motion detectors (*Bewegungsmelder*) everywhere nowadays.
4. Technology is changing so fast that we often have ... courses at work.
5. Alan was lucky to get the job. There were over 40 ... for it.
6. My qualifications are fine but they want at least three years ... . I only have two.
7. Albion Engineering is a big ... of parts to the car industry.

2 **Complete Björn's answer to CSS's ad with the English equivalents of the German words in the box. The numbers in brackets show you roughly where the English words go.**

Anforderungen (11–15)	*Elektriker/in* (6–10)	*Lebenslauf* (11–15)
Antwort (11–15)	*Erfahrung* (11–15)	*qualifiziert* (11–15)
Anzeige (1–5)	*Erfolg* (6–10)	*sicher* (1–5)
Ausbildung (6–10)	*jung* (1–5)	*Stelle* (1–5)
Einzelheiten (6–10)	*Kenntnis* (1–5)	*Wartungs-* (6–10)

Von: Björn Drexler
An: tara.ryan@hr.css.ie
Betreff: Application for ... [1] of installation technician

Dear Ms Ryan
I was very interested to see your ... [2] for installation technicians on the www.jobs.ie website. I would like to apply for such a position as I want to improve my ... [3] of English and get some international experience while I am still ... [4] and single.
I am also ... [5] that I could make a real contribution to your company's ... [6] in Germany.
Here are some more ... [7] about myself: At present, I am working as a ... [8] electrician at Siemens AG in Duisburg. I did a three year ... [9] as an industrial ... [10] there and then took a full-time job with the company. I now have three further years' ... [11]. You can see from this that I am fully ... [12] and fit the ... [13] you give in your advertisement. I attach my ... [14] with this email and I hope for a favourable ... [15].
Yours sincerely
Björn Drexler

3 **What would you write to make the email fit you? Make any necessary changes and add two or three sentences about your personal experience and expectations.**

Björn's CV

Björn Drexler

Personal details

Name	Björn Drexler
Address	Robert-Koch-Straße 30, 45276 Essen, Germany
Date and place of birth	28 May 1984, Essen
Nationality	German
Daytime phone number	+49-203-1991-34
Email address	björn.drexler@okipost.de

Education

2001–2003	Fachoberschule für Technik, Essen (technical college) Fachhochschulreife, Klasse 12 (entry certificate, university of applied science) average grade: 1.8 on 6-point scale from 1 (very good) to 6 (very unsatisfactory)
1994–2000	Willy Brandt Comprehensive School, Essen Abschluss Klasse 10 (school-leaving certificate, equivalent of Irish Junior Certificate) average grade 2.1

Work experience

2000	RWE power generation, Essen (stores)
1999	Achim Stein Elektroinstallationen, Essen (helping on site)

Experience

2006–to date	maintenance electrician, Siemens AG, Duisburg
2003–2006	apprenticeship as electrician, Siemens AG, Duisburg

Skills

since 1997	playing member, Rot-Weiss Essen football club
Languages	English – good (some knowledge of technical English)
IT skills	Word, Excel; Linux; AJAX Software, circuit design
Other	driving licence, EU class C

References	references can be supplied upon request

4 **Give the English equivalent for the following German terms from the CV on page 49.**

1 *Fachoberschule für Technik*
2 *Fachhochschulreife*
3 *Durchschnittsnote*
4 *Gesamtschule*
5 *Schulabschlusszeugnis*
6 *entspricht*
7 *Berufspraktikum*
8 *Lehre*
9 *EDV-Fertigkeiten*
10 *Empfehlungsschreiben*

5 **Use Björn's CV as a model to write a CV for yourself.**
Go to page 6 for a list of common technical occupations.

A job interview

6 **Work with a partner. Read the interview and try to put in the missing words. There are generally clear clues.**
Tr11 **Then listen to the CD to check your answers and fill any gaps.**

Björn Drexler is sitting outside a conference room at the Sheraton Düsseldorf Airport Hotel. Two other people are sitting there as well. They are waiting to see Tara Ryan and Liam O'Brien of Celtic Security Systems Limited. The conference room door opens and Liam looks out ...

Liam Mr Drexler? Could you come in, please? This is Tara Ryan from Human Resources, Mr Drexler, and my ... [1] is Liam O'Brien. I'm in charge of engineering services at CSS.
Björn Good morning. It's nice to meet you.
Liam Well, take a seat, please. Can I give you something to ... [2]? A coffee, perhaps, or something cold?
Björn No, thank you. I'm fine.
Liam Okay, then. Tara?
Tara Well, Mr Drexler, thank you very much for your job ... [3]. First, I'd just like to ask you a few ... [4] about your experience. Now, at present you're working for ... [5], aren't you?
Björn Yes, that's right.
Tara Well, everybody knows Siemens, of course, but can you tell us a bit about your duties there?
Björn I'm an ... [6]. I work in plant maintenance.
Tara Sure, but I'd like to get a picture of what you actually do. I mean, you go to ... [7], clock on ... Then what happens?

Björn Well, I check to see if there are any urgent … [8]. I do them first, of course. Then I get on with the routine jobs.

Tara Such as?

Björn Well, lots of things. Anything from changing light … [9] to servicing and repairing complicated equipment. If it's electric, I do it.

Liam Excuse me, Tara. What about security systems, Mr Drexler? Do you know anything about them?

Björn Well, not much. I service security equipment occasionally, but …

Liam Can you be a little more precise, please, Mr Drexler? What sort of security … [10]?

Björn Well, mainly fire equipment.

Liam I see. So you don't have any experience with the kind of equipment we produce?

Björn No, not really, but it's all electric, isn't it? And there's three months' … [11], isn't there?

Liam True, but … [12] always helps, doesn't it? Now, Mr Drexler, another point. Your qualifications. Right up there at the top of the class. Aren't you overqualified for this job? You can be *too* well … [13], you know.

Björn Look, Mr O'Brien, nobody leaves Siemens without thinking about it very … [14]. I enjoy my work and I try to do a good job. But in spite of my … [15] I'm not going anywhere. Your job offers variety and future prospects. That's what I'm … [16] in.

Liam Good. That's what I wanted to hear. Tara?

Tara Just two questions. You live in Essen, right?

Björn That's right. I live at home with my … [17].

Tara Oh. I didn't know that. You're – um, let's see – 25 and you still … [18] at home? Excuse me, but isn't that a little, well, unusual, Mr Drexler?

Björn Why? I pay my share, of course, but I get on very well with my parents. My mother's a great cook and my father works at Siemens, too. We often go to work together. I don't want to move unless I have to.

Tara Yes, well, perhaps you're right, but … Anyhow, our German associate is in Leipzig. Any … [19] with moving there?

Björn No problems at all. That's where my father comes from. We still have family there.

Tara Fine. And another point. You will work alone. Is that a problem? Will you miss the social life of your present workplace?

Björn I don't think so. And I'll meet lots of new … [20], won't I? That's one of the attractions of the job.

Tara That's true. Well, Mr Drexler, let me just tell you where we go from here …

7 **Choose the correct answer.**

1	First of all, Liam	introduces Björn to Tara. introduces Tara and himself to Björn. offers Björn something to drink.
2	First Tara	asks Björn about his job at Siemens. tells Björn about the job at CSS. thanks Björn for his application.
3	Björn tells Tara that he	mainly services complicated machinery. services and repairs anything that is electric. just does general maintenance.
4	Björn doesn't have any experience with	security equipment. automatic doors. the security equipment that CSS makes.
5	Björn may be overqualified because he	did very well in his exams. has a difficult job at Siemens. got a full-time job at Siemens.
6	Björn must move to Leipzig because	he has family there. CSS's associate is there. his father comes from there.

8 **Read the word list and listen to the dialogue. Then say if the statements are true or false. Correct the five false statements.**

Tr 12

Word list

contractor	*(hier:) Handwerksbetrieb*
choice	*Wahl*
mate	*Kumpel*
way over the top	*sehr übertrieben*
smooth	*glatt*
(to) have everything off pat	*auf alles gleich eine Antwort haben*

1. Tara and Liam are talking about the three applicants.
2. The first person they discuss is Björn Drexler.
3. They reject (*ablehnen*) Peter Becker because he's too old.
4. Liam and Tara agree that Becker is not strong enough for the job.
5. Tara thinks that Björn could find it difficult to work away from home.
6. Liam doesn't agree with Tara about Björn's living at home.
7. Tara's own boys are happy to live at home as long as they can.
8. Liam asked Björn about his experience just to see his reaction.
9. Tara says she likes Björn Drexler more than Markus Schwaner.
10. Both Tara and Liam found Markus Schwaner a bit too good to be true.

9 Do you think Björn should get the job? Why/Why not? Choose reasons for your opinion. What does the class think?

Yes, I do.

- CSS's ad refers to growing sales in the EU. They have a real need for installation technicians.
- Interviewers often make negative comments just to see how the person reacts. They don't really mean anything.
- Björn was trained by Siemens and now he works for them full-time. Getting people of this quality must be good for CSS.
- Björn answers the negative questions well – his point about three months' training, for example.

No, I don't.

- Björn doesn't have any experience of the types of security devices that CSS produces. Liam thinks this is a weakness.
- Björn's obviously very clever. He could become bored with just installing equipment and find a more demanding job. Liam seems to think this, too.
- Björn still lives at home. He's not ready to go out in the field by himself yet. He needs to be more independent. Tara seems to think this, too.
- There's no way that Björn can learn enough in just three months to install complicated equipment on his own.
- Björn says himself that he likes his work at Siemens and nobody should leave the company without a lot of thought. Is he ready to go yet?

I think ...
In my opinion ...
Do you think so? I think ...
Björn should/shouldn't get the job because ...
Tara/Liam is right when she/he says / seems to think ...

! Ich meine = I think, **nicht** ~~I mean~~

Practising grammar

Remember: Simple present → Begleitgrammatik, Abschnitt 1, Seite 150

1 I **live** at home with my parents.
2 Björn often **visits** job sites.
3 Björn **doesn't want** to move unless he has to.
4 **Do you know** anything about them?

1 **Put the verbs into the *simple present*. Be careful about negative statements and questions.**

1 The company ... [1] (make) security equipment and ... [2] (sell) it all over the EU.
2 Björn Drexler ... [1] (live) in Essen but he ... [2] (work) at Siemens in Duisburg.
3 You ... [1] (need) English if you ... [2] (want) to work in the global village.
4 Björn sometimes ... [1] (go) to work with his dad. He ... [2] (be) at Siemens, too.
5 Tara ... [1] (think) it is unusual that Björn ... [2] (not have) a flat of his own.
6 ... (CSS + give) new technicians three months' training?
7 ... (you + think) that Björn will get the job?
8 We ... (not work) on Saturdays.

2 **Use the ideas below to say what you *normally/often/sometimes/never* do – or don't do – in your everyday life.**

argue with my parents/brother/sister	go clubbing/dancing/shopping	play computer/video games
catch the bus	go to LAN parties	send text messages
chill out	go to the movies/to the pub/to the disco	surf the net
come home late	hang around with friends	visit chatrooms
do homework	leave for school/my job	watch TV/a DVD
drink alcohol	listen to CDs/the radio	
email friends		

Example: I never go to the pub during the week.

■ Kurze Adverbien der Häufigkeit wie ***often***, ***never***, ***always*** usw. stehen unmittelbar vor dem Vollverb.

3 **Ask your partner about his or her everyday life. Swap roles at least once.**

Examples: **When** do you leave for school?
How often do you work?
Where do you go shopping?
Do you **go** to LAN parties?

Remember: Present progressive → Begleitgrammatik, Abschnitt 2, Seite 151

1 Björn **is sitting** in the airport hotel at the moment.
2 They **are waiting** to see Tara Ryan and Liam O'Brien.
3 In spite of my qualifications I'**m not going** anywhere.
4 **Is CSS looking** for technicians at present?

4 **Put the verbs into the *present progressive*. Be careful about negative statements and questions.**

1 We ... (look for) installation technicians at the moment.
2 Björn ... [1] (work) in Duisburg at present but he ... [2] (hope) to get a job in an English-speaking country.
3 Where are Tara and Liam? ... (they + interview) job applicants in Germany today?
4 I can't come now. I ... (help) Dad to wash the car.
5 Tara isn't here at the moment. She ... (take) the new technicians on a tour of the plant.
6 CSS ... [1] (grow) very fast at the moment. The company ... [2] (sell) security equipment all over the EU.
7 CSS ... (offer) a "generous" salary but the job ad doesn't say exactly how much.
8 I ... [1] (not drive) to work today. My car ... [2] (be) serviced.

5 **Use the material to say what is happening in the drawings.**

1 mechanic/ change/tyre

2 apprentice/ wash/ car

3 electricians/ repair/car lights

4 mechanics/ install/engine

5 mechanic/ check/oil

6 apprentice/ fill/fuel tank

7 apprentices/ clean/workshop

8 apprentices/ put away/tools

Remember: Kurzantworten		→ Begleitgrammatik, Abschnitt 3, Seite 151
Frage	**Kurzantwort**	
	bejaht	**verneint**
Does Björn work in Duisburg?	Yes, he does.	No, he doesn't.
Did you answer the job ad?	Yes, I did.	No, I didn't.
Is he a mechanic?	Yes, he is.	No, he isn't.
Are you interested in the job?	Yes, I am.	No, I'm not.
Are Tara and Liam interviewing people?	Yes, they are.	No, they aren't.
Can Tara speak German?	Yes, she can.	No, she can't.

6 **Put in short answers. Look at the examples first.**

Examples:
- Does Björn live in Duisburg? (✘) *No, he doesn't.*
- Is Björn driving to work today? (✔) *Yes, he is.*
- Are Anton and Christine Austrian? (✔) *Yes, they are.*

1. Is CSS looking for installation technicians? (✔)
2. Are the applicants waiting in the conference room? (✘)
3. Did Björn go for an interview yesterday? (✔)
4. Can Tara and Liam understand German? (✘)
5. Does Björn's father come from Essen? (✘)
6. Did you find the job on a website? (✔)
7. Do you often play computer games? (✔)
8. Is Liam in charge of human resources? (✘)
9. Does Björn have any problems about moving? (✔)
10. Am I next? (✘)

Remember: Frageanhängsel		→ Begleitgrammatik, Abschnitt 4, Seite 152	
Aussage	**+ / –**	**Frageanhängsel**	**+ / –**
You **are** working for Siemens,	+	**aren't** you?	–
Björn **can't** speak French,	–	**can** he?	+
They **haven't** found a flat yet,	–	**have** they?	+
There **is** three months training,	+	**isn't** there?	–
Experience always **helps**,	+	**doesn't** it?	–
Peter **didn't** have the right experience,	–	**did** he?	+
The trainees **must** spend three months in Dublin,	+	**mustn't** they?	–
You **won't** be at the meeting tomorrow,	–	**will** you?	+

7 **Complete the sentences with the missing question tags. Look at the examples first.**

Examples:
- Björn still lives at home, ... *doesn't he?*
- Markus and Peter didn't get a job, ... *did they?*
- Tara and Liam are Irish, ... *aren't they?*
- Liam must leave for the airport at 3, ... *mustn't he?*

1 CSS is an Irish company, ...
2 Two applicants are from Austria, ...
3 Björn didn't have an interview in Dublin, ...
4 The interview wasn't in German, ...
5 Anton doesn't come from Germany, ...
6 Björn didn't find the job ad in a newspaper, ...
7 Björn's father works at Siemens, too, ...
8 Liam O'Brien will be Björn's boss, ...
9 At his age, Björn should get a flat of his own, ...
10 The company hasn't found an associate in Vienna yet, ...

8 **Work with a partner. Person A asks a question about the highlighted part of the sentences. Then person B answers it. Swap roles at least once.**

Examples:
- Björn Drexler wants to get a job with an Irish firm. (Who ...?)
  A *Who wants to get a job with an Irish firm?*
  B *Björn Drexler.*
- Björn lives in Essen. (Where ...?)
  A *Where does Björn live?*
  B *In Essen.*

1 Björn can speak good English. (Who ...?)
2 The three men are at the Sheraton for an interview. (Why ...?)
3 Tara and Liam are interviewing three technicians. (How many ...?)
4 Björn Drexler does not know much about security devices. (Who ...?)
5 Peter Schwaner is studying engineering. (What ...?)
6 Tara thinks it's unusual that Björn still lives at home. (What ...?)
7 Björn found the job ad on the internet. (Where ...?)
8 Björn reads the job ads in the newspaper on Saturdays. (When ...?)
9 CSS has associates in 24 countries. (In how many ...?)
10 Björn wants to work for an English-speaking firm to improve his English. (Why ...?)

Warm-up

Tr 13 **These people are travelling from Britain to the continent. Listen to what they say and complete the speech bubbles with the missing details. All the words belong to the wordfield 'travel'.**

1 Hello. My name's Tom Ross. Well, I'm afraid I get ... [1] very easily so I don't like the ... [2]. The Eurotunnel from the ... [3] near Folkestone to Calais is just the thing for me. With a ... [4], it costs about €150 return, depending on when you ... [5]. It's very quick, too. The ... [6] is only about 35 minutes each way.

2 Hi. I'm Jodie Harris from Bristol. I don't like ... [1]. The Eurotunnel is about 50 kilometres long and it's nearly all under the ... [2]. I much prefer the car ferries. ... [3] are quite cheap and the ... [4] are really comfortable. I think the Dover-Dunkerque ... [5] is best. That costs about €80 for a car with a ... [6] and up to four ... [7]. It's quite quick, too – about one and three-quarter hours.

3 My name's Ann Morgan. Well, in my opinion ... [1] are the answer. They're so cheap and quick. We ... [2] from Liverpool to Cologne with easyJet last year for less than €20 each, though we had to pay ... [3], of course. Then we ... [4] a car at cheap weekend ... [5] or took a bus or ... [6]. For example, a VW Polo cost us €70 for two days and we ... [7] up the Rhine. That was great!

1 **Use the material to describe Björn's route by car from Essen to Dublin.**

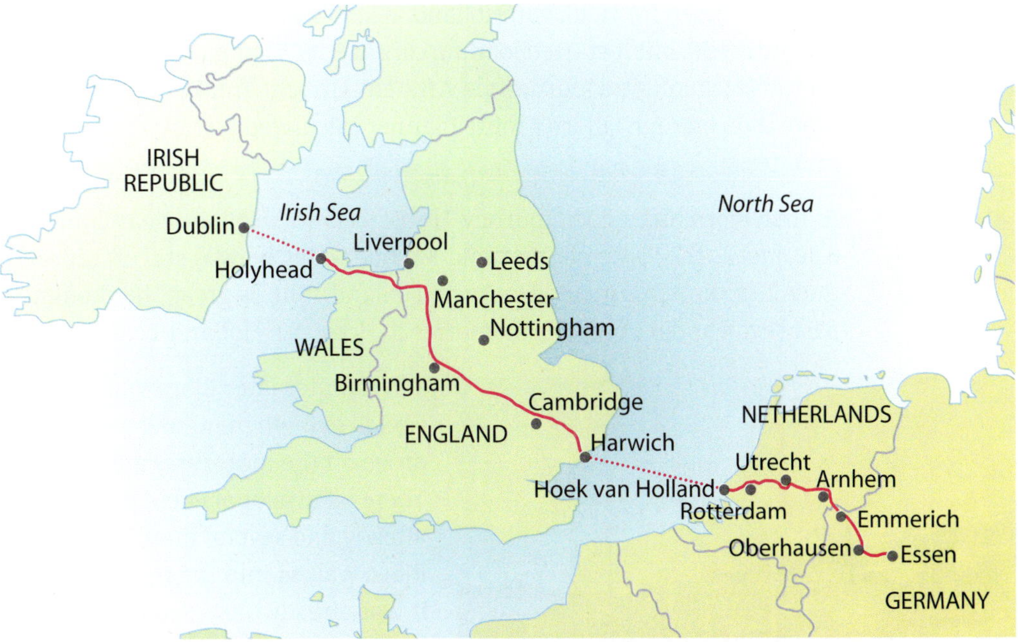

begin one's journey in ... go from ... to ... travel to ... via ... and ... take the by-pass round Arnhem/...	cross the border into ... at ... travel through ... countries
have a break at/near ... buy fuel at/near ...	take a ferry across the ... from ... to ... cross the North Sea/Irish Sea board a ferry at ... ferry port land in .../at ... ferry port

2 **Read the text and put in the missing prepositions from the box.**

across (3x) at by for from (2x) in (3x) on (2x) to (6x) with

Björn wants to travel ... [1] Essen ... [2] Dublin ... [3] car because he is taking a lot of luggage ... [4] him. Hc plans to leave his home ... [5] Essen ... [6] Dublin ... [7] Friday, 27 April. First, he will drive ... [8] the Netherlands ... [9] the ferry port ... [10] Hoek van Holland. There, he will travel ... [11] a fast ferry ... [12] Harwich ... [13] eastern England. Then he will drive right ... [14] England ... [15] Holyhead ... [16] Wales. Finally, he will catch a ferry ... [17] the Irish Sea ... [18] Dublin. Tara says this route is better than the one Björn originally chose ... [19] Calais ... [20] Dover because it avoids London.

3 **You are Björn Drexler. Write an email to Tara Ryan telling her about your travel plans and when you expect to arrive in Dublin.**

Tr 14

Björn's journey to Dublin

Björn left Essen for Hoek van Holland at 5.30 a.m. on Friday, 27 April. The ferry didn't sail until 9.30, but it started loading cars at 9.00. That gave him three and a half hours for the 260 kilometre journey via Arnhem, Utrecht and Rotterdam. Surely that was more than enough? Björn's route planner calculated just two hours and thirty-three minutes.

At first, Björn had a good journey. He crossed into Holland near Emmerich at about 6.30, just as the route planner said. An hour later, he was still on time as he went on to the A20 to Rotterdam and Hoek van Holland. "This is great," he thought. "It's only half past seven and it's only 50 kilometres to Hoek van Holland. I can stop for coffee soon."

Ten minutes later Björn was driving towards the Gouda exit when he saw a restaurant sign. He drove off the motorway and into the car park. There were a lot of cars, trucks and vans there already and several more were following him. Björn walked into the restaurant and he saw that it was already busy. People were having breakfast and long queues were standing at the coffee machine and the cash desk. When Björn joined the coffee queue five or six people were waiting in front of him, and a man at the machine was very slow. "Come on," Björn thought. "I don't have all day."

At last Björn got his coffee and found a seat at a crowded table. People were chatting and reading newspapers. Björn looked at the clock on his mobile. It was nearly 8 o'clock. "Relax," he thought. After all, the ferry didn't even start loading for an hour and it was now less than 50 kilometres to the port. He finished his coffee and walked out of the restaurant into the car park. He looked towards the motorway. There was heavy traffic and it wasn't moving very fast. Stop and go. There was a long queue at the access ramp, too, and the filter was operating. It only let one vehicle on to the motorway at a time.

Half an hour later, Björn was sitting in a tailback in the rush-hour traffic around Rotterdam. "My God," he thought. "Why did I stop for coffee? What a fool! I've only done about ten kilometres." He saw a sign at the side of the road. It said "Hoek van Holland 37". Björn looked at the clock on the dashboard. 08.39, it said, and flicked over to 08.40.

At last the queue of traffic started to move. The automatic traffic management system around Rotterdam gave a maximum speed of 40 km/h, but a lot of drivers were going much faster. The result? Another tailback with more stop and go. It was already past 9 o'clock when a police car and an ambulance overtook the queue. They were travelling fast along the hard shoulder in Björn's direction. Their blue lights were flashing and their sirens were screaming urgently. A few minutes later, the warning lights on the cars in front of Björn started to flash and all traffic stopped. Björn looked up the road. Some people were already getting out of their cars. A few were using their mobile phones. "There's an accident ahead," Björn thought. "Oh, well, that's it. There's no way I can get the nine-thirty ferry now. I'll have to take a later one."

The date (British English)

We write:	**We say:**
August 2nd 2007	August **the** second, two thousand and seven
10th May 2008	**the** tenth **of** May, two thousand and eight
8. September 2010	**the** eighth of September, twenty ten

Note how to say the years 2000 to 2009, and then 2010 and later.

4 **Use the *simple present* to say what role the following people and places play in the text. Give as much information as you can. Look at the example first.**

Example: – Hoek van Holland *This is the ferry port in the Netherlands where Björn takes a ferry to Harwich in England.*

1 Emmerich
2 A20
3 the Gouda exit
4 a man at the coffee machine
5 a filter at an access ramp
6 tailbacks
7 a lot of drivers
8 a police car and an ambulance
9 warning lights
10 mobile phones

5 **Work with a partner. Use the prompts to ask each other questions about Björn's journey. Look at the examples first.**

Examples:
– when/Björn/leave?
When did Björn leave (Essen for Hoek van Holland)?
At five thirty on Friday, the twenty-seventh of April.
– how far/Essen/ Hoek van Holland?
How far is it from Essen to Hoek van Holland?
It's two hundred and sixty kilometres (via Arnhem).

1 how long/route planner/say/journey/take?
2 where/Björn/stop?
3 why/take so long/restaurant?
4 why/Björn/feel uneasy (*unbehaglich*)/car park?
5 when/Björn/realise/not catch?
6 what/Björn/decide/do?

6 **Björn finally arrives at Hoek van Holland at 10 o'clock. He parks his car and walks to the ferry terminal. There he finds details of the next ferry to Harwich – and what that means for his continued journey to Holyhead and Dublin.**

Hoek van Holland to Harwich

From **£49** Car & Driver

Single motorist fares

Prices (£) shown are for Economy motorist fares for car or MPV under 6m long and up to 1.8m high plus driver. For return fares add the appropriate single fares together.
Only Economy and Flexi fares are available on this route. No Service 25/26 December.

Hoek van Holland – Harwich				
Craft	**Sailing type**	**Sailing time**	**Arrival time**	**Economy fares from:**
Superferry	Day	09:30	15:00	49
Superferry	Day	14:30	20:00	49
Superferry	Overnight	22:00	06:30	49

Harwich – Hoek van Holland				
Craft	**Sailing type**	**Sailing time**	**Arrival time**	**Economy fares from:**
Superferry	Day	09:00	16:15	49
Superferry	Overnight	23:45	07:45	49

a) **Work with a partner. First, use the timetables and the travelling times below to revise Björn's travel arrangements from Hoek van Holland to Dublin. Add any further details such as rest periods yourselves.**

Travelling times	
Hoek van Holland – Harwich	Day: 6 hrs 30 mins + ca. 30 mins loading/off-loading Night: 9 hrs 30 mins + ca. 30 mins loading/off-loading
Harwich – Holyhead	Allow at least 7 ½ hours by fastest route
Holyhead – Dublin	1 hr 50 mins + ca. 20 mins loading/off-loading

b) **One of you is Björn and the other is Tara or her assistant, Kevin. Use the German prompts to write and act out a telephone dialogue about Björn's situation.**

Björn:
sagt, was passiert ist
berichtet über die nächste Fähre
sagt, dass er auch die vorgesehene Holyhead-Fähre verpassen wird
sagt, dass er (u.U.) irgendwo in England übernachten muss oder vielleicht durchfährt
bittet um Rat
entschuldigt sich für die Schwierigkeiten
verabschiedet sich

Tara/Kevin:
fragt, wie/warum Björn die Fähre verpasst hat
reagiert locker: kann passieren, Pech
bittet Björn, die nächstmögliche Fähre zu nehmen
schlägt Björn vor, dass er irgendwo in England übernachtet. Soll keine Risiken eingehen
sagt, dass er/sie das Hotel in Dublin über die verspätete Ankunft benachrichtigen wird
reagiert mit „Keine Ursache"
verabschiedet sich mit „Weiterhin gute Fahrt"

Practising grammar

Remember: Simple past	→ Begleitgrammatik, Abschnitt 5, Seite 152
Beispielsatz	**Form**
1 Björn **crossed** the Dutch border near Emmerich.	Infinitiv + *-ed*
2 He **left** Essen at half past five.	unregelmäßig, s. Seite 162
3 The ferry **did not/didn't leave** until 9.30.	*did not/didn't* + Infinitiv
4 **Did Björn forget** the roadworks near Rotterdam?	*did* + Subjekt + Infinitiv

1 **Complete the sentences by replacing the German verbs in brackets with their English equivalents from the box.**

bought	didn't obey	lost	There were	wanted
Did Tara warn	drove	missed	told	weren't
didn't cross	heard	stopped	took	

1 Björn ... his ferry tickets to England and Ireland online. (*kaufte*)
2 He ... 1 for coffee at a restaurant but he ... 2 a lot of time there. (*hielt an, verlor*)
3 Several other passengers ... their ferry because of heavy traffic. (*verpassten*)
4 ... Björn about the roadworks in Holland? (*Warnte Tara*)
5 Björn ... 1 to Dover because he ... 2 to avoid London. (*überquerte [den Kanal] nicht, wollte*)
6 ... 1 bad tailbacks because drivers ... 2 the speed limit. (*Es gab, befolgten nicht*)
7 Björn ... 1 a later ferry and then ... 2 to Holyhead as fast as possible. (*nahm, fuhr*)
8 Tara and Kevin ... 1 annoyed when they ... 2 Björn's news. They ... 3 him not to worry about it. (*waren nicht, hörten, sagten*)

2 **It is Friday, 10 November. Read the extract from a car repair workshop diary for 9 November. Then use the *simple past* to write sentences about what the people did.**

Thursday, 9 November

Skoda Octavia, change oil (renew all filters), Roy
VW Fox, fit winter tyres, Ravi
Renault Clio, replace battery, Susan
Land Rover, adjust brakes/steering, Brian/Eddie
Ford Escort, check anti-freeze & fill up as necessary, Ben
Collect Rover 25, 10 Mill Lane, 10 o'clock, Sharon/Jack
BMW, wash/polish, Dave/Joe
VW Golf, repair rear windscreen wiper, Charlie
Order spare parts for Vaneo, Barbara
Carry out road test, Fiat Punto, Mike

Examples:

Roy changed the oil and renewed all filters on a Skoda Octavia.

Ravi fitted winter tyres to a VW Fox.

Susan ...

Remember: Past progressive	→ Begleitgrammatik, Abschnitt 6, Seite 153
Beispielsatz	**Form**
1 Björn **was driving** towards Gouda when he saw a restaurant sign.	*was/were* + *-ing*-Form
2 At 08.40, traffic was heavy and it **was not/ wasn't moving** very fast.	*was/were* + *not* + *-ing*-Form
3 **Were** drivers **obeying** the speed limit near Rotterdam?	*was/were* + Subjekt + *-ing*-Form

3 **Look at the workshop diary again. What were the mechanics doing at 10 o'clock last Thursday? Form sentences with the correct form of the *past progressive*.**

1 Roy *was changing* the oil on a Skoda Octavia.
2 Ravi ... to a VW Fox.
3 Susan ... on a Renault Clio.
4 Brian and Eddie were ...
5 Ben ...
6 Susan and Jack ...
7 Dave and Joe ...
8 Barbara was in the office. She ...

4 **Form negative statements and questions with the *past progressive*.**

1 ... (Roy and Ben/work) outside when it started to rain?
2 Susan ... (not do) anything at 11 o'clock. She was in the canteen.
3 We ... (not repair) a BMW yesterday. It was a Ford.
4 ... (Barbara/telephone) when you went into the office?
5 ... (you/surf) on the internet when I spoke to you?
6 I ... (not replace) a battery when the accident happened.

5 **Use the *simple past* or the *past progressive* to complete this accident report from Expatica, an English-language Dutch newspaper.**

Rotterdam, Friday – A police officer *told* [1] journalists that they ... [2] (be) unable to say what ... [3] (cause) the accident on the A20 near Rotterdam on Friday morning that ... [4] (test) the nerves of hundreds of people on their way to work. Traffic was *not moving* [5] fast in the morning rush hour, and road conditions ... [6] (be) good. Although the sun ... [7] (shine) at the time, it ... [8] (not dazzle) (*blenden*) drivers who ... [9] (travel) west. The driver who ... [10] (cause) the accident ... [11] (not use) a mobile phone. Alcohol or drugs also ... [12] (play) no part in the accident. Why, then, did he ... [13] (move) into the fast lane without warning and ... [14] (drive) into the side of a van that ... [15] (overtake) him? Neither driver ... [16] (be) hurt in the accident, but it ... [17] (set off) a chain of events that ... [18] (block) the motorway for over an hour. Unfortunately, two cars that ... [19] (follow) the van ... [20] (not stop) in time. According to witnesses, both cars ... [21] (travel) far too fast for the traffic conditions and ... [22] (not have) enough space between them.

Warm-up

Björn spent his first week in Dublin in a small hotel near the airport. The hotel was fine but it was noisy and very expensive – €85 a night without breakfast.

Björn decided to find a flat as soon as possible.

1 Look carefully at all the ads. Write these abbreviations out in full. They are in the same order as in the text.

1 furn.	**3** liv-rm.	**5** bathrm.	**7** p.w.	**9** gdn.
2 h&c	**4** kit.	**6** sep.	**8** bedrm.	**10** dep.

1

ASHTOWN
Compact furn. bedsitter with h&c water in bungalow. Shared bathroom/WC. Use of garden. Near university. Ideal for singles. €100 per week, no deposit. Call 788432 after 6 p.m. (weekdays only).

2

CHERRY ORCHARD
Fully-furn. 1-bed flat, liv-rm. with kit., new bathrm./WC. Above shop with sep. entrance. Singles, one-parent families welcome. €180 p.w. Call 671200 evenings, weekends.

3

GLASREVIN
Large 1st floor flat in converted Victorian house. Fully furnished. Large liv-rm., kit, 3 bedrms. Use of garden, garage. €250 p.w. Call 893274 any time.

4

DUBLIN CENTRE
Attractive apartment in new building, built-in compact kitchen with breakfast bar, luxury bathroom, lift, parking. Short lets possible. €200 p.w., deposit one month's rent. Call Sara Mills on 296165, office hours.

5

SUTTON EAST
Small unfurnished 2-bed cottage near water. Living-room, fitted kitchen, ground-floor bathrm. Small gdn. Residents' parking. Ideal for older couple. No pets. €300 p.w., €1000 dep. Tel: 7243567

6

Fully renovated HOUSEBOAT
on Liffey. Large living area, 2 beds, fitted kitchen, bath/WC. All mains services. Parking. Ideal for young couple. No short lets. €300 p.w. + taxes. **Call Sean Doyle, 672918 any time.**

A

B

C

D

E

F

G

H

2 Match the ads 1–6 with the photos A–H. Give a reason for your choice. Then think up similar ads for the two photos that don't fit.

- I think the flat/bedsitter in the first/second/... ad is in house/cottage/... A/... In the ad it says ...
- Photo ... shows the house in the first/second/... ad.
- Photo ... doesn't fit/illustrate any of the ads.

3 Which accommodation would you choose? Say why.

- I'd choose ...
- I definitely wouldn't choose ...

modern/new; too big/small; too expensive; too noisy; too much work; fully furnished; have all modern conveniences (*allen Komfort haben*)/good parking/a garage/...; not have to pay a deposit

Tr 15

Björn's first day

It's Monday, 30 April – Björn's first day at work. He leaves his hotel in Merrion Road at 8.30 and drives to Celtic Security System's offices in Jamestown Business Park. Traffic is heavy and the journey takes nearly 40 minutes, much longer than Björn expects. He parks his car and walks to the main entrance of the office building. He looks at his watch. It's already 9.15. "Oh, my God," he thinks. "This is terrible. I'm 15 minutes late."

1 **Work with a partner. First read the text and try to fill in the missing words. There is often a clue to help you to do this. Then listen to the CD to check your answers and fill in any empty gaps.**

Carol Good morning. Can I … [1] you?

Björn Good morning. My name's Björn Drexler. Can I … [2] to Ms Ryan, please? I had an appointment for 9 o'clock, but I'm a bit … [3], I'm afraid.

Carol Oh, pleased … [4], Björn. I'm Carol. Tara told me to expect you. Just one … [5], please. I'll tell her that you're here.

Tara Hello, Björn. How … [6]? Welcome to CSS.

Björn Hello, Tara. Look, I'm … [7] I'm late. It took much longer to get here than I thought. First I miss the … [8] and now this. You must think I'm a real …

Tara Don't … [9], Björn. The … [10] is terrible in Dublin. I was a bit late myself this morning. Anyhow, let's go up to my … [11] and I'll tell you what we have planned for today.

Björn Fine – and thanks.

Tara Well, take a seat, Björn. … [12]?

Björn Sure. I'd like a coffee.

Tara Oh, before I forget – here's your name tag. Please … [13] it at all times when you're in the plant. You know, security …

Björn OK.

Tara At 10 o'clock, we'll go to the conference room and I'll … [14] you to the three other trainee installation … [15]. Then I'd like to do some administrative stuff with you all. Insurance, tax, that sort of thing.

Björn Oh, I have a … [16] there, Tara. What about my … [17]? It's registered in Germany. Do I need to do anything about that?

Tara No … [18], Björn. You can drive your car on German number plates for six … [19], you see, and you are only here for three.

Björn I see. Fine.

Tara Anyhow, the … [20] work won't take more than 30 minutes or so and then Liam O'Brien will join us at about … [21]. He will go through the training course with you, which will take about 45 minutes. Then after a 15-minute break for coffee, Liam will take you on a … [22] of the plant. That will begin at about … [23] and take around 90 minutes. Then we all meet in the canteen for lunch at 1 o'clock. Any … [24] so far, Björn?

Björn No. That sounds fine.

Tara First …

2 **Give the questions that produced these answers. Use the question word in brackets. Look at the examples first.**

Examples:
– Björn's hotel is in Merrion Road. (where)
Where's Björn's hotel?

– He started work on Monday, 30 April. (when)
When did he start work?

1 He left for Jamestown Business Park at 8.30. (when)
2 He took nearly 40 minutes to reach CSS's offices. (how long)
3 The receptionist's name is Carol Brennan. (what)
4 Björn asked to speak to Ms Ryan. (who)
5 Björn's appointment was at 9 o'clock. (when)
6 Björn was late because of the heavy traffic. (why)
7 All together, there are four trainee technicians. (how many)
8 Björn must always wear his name tag for security reasons. (why)
9 Björn will meet the other technicians in the conference room. (where)
10 Liam wants to talk to the trainees about the training course. (what)

3 **a) Give the English equivalents of these German words and phrases. They are in the same order as in the text.**

1 *Gewerbegebiet* (1–5)
2 *Verkehr* (1–5)
3 *Fahrt* (1–5)
4 *Haupteingang* (1–5)
5 *Termin* (6–10)
6 *erwarten* (6–10)
7 *tut mir leid* (11–15)
8 *Machen Sie sich keine Gedanken!* (11–15)
9 *fürchterlich* (11–15)
10 *Namensschild* (16–20)
11 *Versicherung* (21–25)
12 *Nummernschilder* (26–30)
13 sich *anschließen* (31–35)
14 *Lehrgang* (31–35)

b) Use some of the English words you have found in a) to complete the sentences.

1 This gate is for deliveries. The ... ... to the plant is in Harbour Road.
2 The town is trying to attract firms by building a modern ... ... near the airport.
3 Patient: What's going on? It's 10 o'clock now and my ... with Dr Brown was for 9.
4 Tom can't come to the meeting but he hopes to ... us for lunch.
5 Security guard: I'm sorry. You can't come into this building without a ... ...
6 The other driver in the accident didn't have any ... so I had to pay for the repairs.
7 We are still planning the project at the moment. The ... ... starts in May.
8 How long did the ... from Essen to Hoek van Holland take?

Begrüßung und Verabschiedung		
Pleased to meet you. How are you? How are you doing?	*Freut mich, Sie/Dich kennen zu lernen.* *Wie geht es Ihnen/dir?* *Wie geht es Ihnen/dir?*	
→ Die gängige Antwort auf **How are you?** lautet **Fine, thanks. And you?** oder **I'm fine, too, thanks.**		
Good morning Good afternoon Good evening	*Guten Morgen, Guten Tag* *Guten Tag* *Guten Abend*	0–12 Uhr 12–18 Uhr 18–0 Uhr
Good night	*Gute Nacht*	
Goodbye Cheerio Bye now	*Auf Wiedersehen, Auf Wiederhören* *Tschüs, Servus* *Tschüs, Servus*	
See you later	*Bis nachher*	

4 First look at the situations in a) and say what expressions you would use in English. Then replace the German parts of the dialogues in b) with English and practise the English dialogues with a partner. Use your own names.

a) Situations

1 Sie lernen jemanden kennen. Sie möchten sich namentlich vorstellen und die Person begrüßen.

2 Sie begegnen jemandem, den Sie schon kennen, nach einiger Zeit wieder. Sie möchten die Person begrüßen.

3 Es ist 15 Uhr. Sie möchten eine Verkäuferin begrüßen und sich nachher verabschieden.

4 Sie möchten einen guten Bekannten verabschieden, dem Sie später nochmals begegnen werden.

b) Dialogues

1 A This is ..., ... You haven't met before, have you?
B No. *Freut mich, Sie kennen zu lernen, ...*

2 A Hello, ... *Wie geht es Ihnen?*
B Fine, thanks. And you?
A *Oh, auch gut, danke.*

3 A Good morning. *Mein Name ist ... .* Can I speak to ..., please?
B Good morning, ... *Einen Moment, bitte.* I'll see if he/she is in his/her office.
A *Danke schön.*

4 A Hello, ... *Tut mir leid, dass ich mich etwas verspätet habe.*
B *Machen Sie sich keine Gedanken.*

5 A Well, I must go now or I'll be late for my train. *Auf Wiedersehen.*
B *Tschüs/Servus, ... Bis nächste Woche.*

Tr 16 Finding a flat in Dublin

There were three other trainee technicians in the conference room: Carlos Carrera from Barcelona, Pierre Leclerc from Lyon and Anton Gruber from Graz. Tara introduced Björn to them. "This is Björn Drexler from Germany," she said. "Björn has just met the awful Dublin traffic." Tara pointed to a seat next to Anton. "Perhaps you'd like to sit with Anton, Björn," she said.

Björn and Anton got on together from the start. Anton told Björn that he wanted to rent a flat, too, but good flats were very expensive in Dublin. "And you can forget the cheaper ones," he said. "I have looked at several and they're, well, terrible."

Björn went onto the internet and looked at the ads in local newspapers and even supermarkets, but he was always disappointed. There were very few flats or bedsitters to rent on the market and they were much more expensive than in Germany. "Anton was right," he thought. "The good flats are much too expensive and the cheap ones are too awful to live in."

About two weeks later Björn was sitting in the staff canteen with Tara Ryan. "Have you found a flat yet, Björn?" she asked. "No," Björn said. "I haven't had any luck so far. There aren't many available and everything's so expensive." "I know," Tara said. "And there are so few flats on the market that the letting agencies charge a fortune, of course. It's because of the boom. Ireland's growing much too fast. What about Anton? Has he found anything yet?" "No, he hasn't found anything yet, either." "Look," Tara said, "have you thought of sharing – or caring, perhaps?" "Caring? What's that?" "Well, people are getting older. You get a flat in an older couple's house in return for helping them. You know, taking out the rubbish, looking after the garden, that kind of thing." "No way," said Björn. "I'm only here for three months and I want to concentrate on my work. But sharing is a good idea." "Yes. You and Anton, you seem to get on well together. Why don't you ask him what he thinks?"

5 Read the text and say if these statements are true (T), false (F) or you don't know because the answer is not in the text (O). Correct the wrong statements.

1 With Björn, there were four trainees in the conference room.
2 Two of the trainees came from German-speaking countries.
3 Carlos Carrera comes from Bilbao.
4 Björn didn't like the French trainee at first.
5 Anton told Björn that he was looking for a flat, too.
6 Anton was staying at a hotel in Essex Road.
7 Anton has already looked at some flats but he hasn't found anything yet.
8 Björn didn't find many ads for flats on the internet or in local papers.
9 Björn thought the flat he looked at in Blackrock was far too expensive.
10 Tara and Björn met in the canteen about two weeks later.
11 Tara said it was hard to find a flat in Dublin because of the university.
12 Tara thinks Björn should take a flat in an old person's house.

6

Find words in the text that fit these definitions. They are in the same order as in the text. Then write five sentences with five of the words you have found.

1 person who a firm is training to do a special job
2 room in firm used for meetings, conferences, etc
3 another adjective for 'terrible', 'very bad'
4 verb phrase for 'be friendly with one another'
5 opposite of 'cheap'
6 adjective for 'three/four/...'
7 one room for living, sleeping and cooking
8 firm's restaurant for its workers
9 opposite of 'slump', 'recession'
10 verb meaning *gemeinsam benutzen*

7

Talking

a) Use the outlines and material below to give a 2-minute talk on *either* the advantages (pros) *or* the disadvantages (cons) of sharing a flat with one or more other people.

Pros	🙂
Opening	Say what flat sharing is/involves
Development	List advantages: share rent/running cost, share chores, rent a bigger flat, live in a nicer part of town, ...
Conclusion	Restate your position: have a private room, think something is a (very) good idea, reduce cost of living, always have companionship, ...
Cons	🙁
Opening	Say what flat sharing is/involves
Development	List disadvantages: lose independence/privacy, must always consider (*berücksichtigen*) another person, have problems inviting people, ...
Conclusion	Restate your position: I would never share ..., I cannot imagine ever sharing ..., value one's independence/privacy too much, ...

b) What do other people in the group think? What is the majority opinion and the most popular reason for it?

Six/... other people in the group agree/don't agree with me/us.
The most popular reason for sharing/not sharing is ...

How to take notes while listening	
1	Use **key words** to note down the information you want, never whole sentences.
2	If possible, use headings to arrange your notes systematically.
3	Use **symbols** (=: *is the same as*, +: *and*, %: *per cent*, etc.), **abbreviations** (e.g.: for example, i.e.: *that is*, km: *kilometre*, GB: *Great Britain*, etc.) and **short forms** (dev: *development*, eco: *ecological*, govt: *government*, etc.) to save time while writing.
4	Make a clean copy of your notes as soon as you can – otherwise you may not be able to read your own writing or understand your own abbreviations or short forms.

8 Telephoning: Answering ads

Tr17

Read through the statements and listen to the CD. Take notes. Then complete the statements.

a)
1 Patrick Flynn's phone number is Dublin ...
2 Björn is answering an ad in the ... ...
3 The flat is in ... ..., but Björn should park in ... ...
4 The flat is on the ... floor at number ...
5 Björn has ... questions.
6 The rent is €... per ...
7 Björn and Anton must pay a deposit of ...
8 Björn and Anton want to look at the flat at ...

b) Listen for the English equivalents of these German sentences in the dialogue.

1 *Wer ist am Apparat, bitte?*
2 *Ich rufe wegen Ihrer Anzeige an.*
3 *Bleiben Sie bitte am Apparat.*
4 *Es dauert keine Minute.*
5 *Sie ist noch frei.*
6 *Ich habe nur ein paar Fragen.*
7 *... wenn Sie nichts dagegen haben.*
8 *Das geht in Ordnung.*

c) Work with a partner. Look at the ads for a used car and a used motorcycle below. Write and act out a similar dialogue about one of them. Use some of the language in b).

FORD FOCUS ESTATE TDCi 1800 (Ghia), silver metallic, first registered 2003, 55,000 miles, MoT certificate, good condition, £4,800, Tel 01523 6297815 evenings/weekends, ask for Bob Hunter.

2004 SUZUKI TL1000S
m/cycle, red, grab handles,
perfect condition,
39,800 miles,
£2,700 or near offer,
Mary Bailey,
07879 529888 any time.

Practising grammar

Remember: Future with **will** → Begleitgrammatik, Abschnitt 7, Seite 153

1 I**'ll/will tell** Tara that you are here.
2 We**'ll/will go** to the conference room at 10.
3 The administrative work **won't/will not take** more than 30 minutes.
4 **Will** Björn **be** late for his appointment?

1 **Use the *will-future*. Be careful about negatives and questions.**

1 You ... (find) that accommodation is very expensive in Dublin.
2 You ... (not get) a comfortable flat for less than € 800 a month.
3 ... [1] (you/share) a flat with me, Anton? It ... [2] (be) much cheaper.
4 Anton: Do you think Carol ... [1] (come) out with me, Björn?
Björn: Well, she ... [2] (not go) out with you if you don't ask her, will she?
5 We must go now or we ... (be) late for the meeting.
6 I'm sorry, Liam, but Sally ... [1] (not finish) the report by three o'clock. It ... [2] (take) her much longer than that.
7 ... [1] (we/arrive) by 9? It's a holiday tomorrow and the traffic ... [2] (be) awful.
8 I know one thing. I ... [1] (not go) to work by car again. I ... [2] (borrow) a bike!

Remember: Present perfect → Begleitgrammatik, Abschnitt 8, Seite 153

1 Tara? It's Carol. Björn Drexler **has just arrived**.
2 Björn **has worked** for Siemens **since 2006/for three years**.
3 Björn and Anton **haven't/have not found** a flat **yet**.
4 **Have** the technicians **finished** the tour of the plant?

2 **Use the *present perfect*. Be careful about negatives and questions.**

1 Tara, Björn ... [1] (just/phone) to say that they ... [2] (find) a flat.
2 ... (Björn and Anton/decide) to share a flat?
3 I ... (not buy) a bike yet, but I hope to find one soon.
4 I ... [1] (not learn) a thing yet. I ... [2] (hear) all this many times before.
5 ... (you/meet) the other trainees, Björn?
6 We ... [1] (not live) in Dublin long. We ... [2] (be) here for only about two weeks.

3 **Use the *present perfect* to say what has just happened. Add any missing words.**

1 plane/land
 – *The plane has just landed.*
2 Ben/buy/paper
3 Carol/write/email
4 Tom/switch off/computer
5 Sally/sell/scooter
6 Josh and Mark/take/dog/walk

4 **Translate the sentences into English. Use the *present perfect*.**
Look at the information in the box first.

I have lived in Dublin **since May**. *Ich wohne seit Mai in Dublin.*	since + **Zeitpunkt**
He has worked here **for three years**. *Er arbeitet seit drei Jahren hier.*	for + **Zeitdauer**

1 Björn arbeitet seit zwei Monaten bei CSS.
2 Anton kennt Björn seit seinem ersten Arbeitstag.
3 Tara wohnt in Dublin, seit sie mit der Schule fertig ist (*left school*).
4 Enrico ist gerade nach Hause gegangen.
5 Hast du Anton Gruber heute schon gesehen?
6 Anton und Björn haben immer noch (*still*) keine Wohnung gefunden.
7 Ich habe gerade ein gebrauchtes Fahrrad gekauft.
8 Wir haben diesen Film schon zweimal gesehen.

Remember: Connectors

1 The traffic is heavy **and** the journey takes longer **than** Björn expects.
2 Björn's hotel was fine **but** it was noisy.
3 Tara was very friendly **although** Björn was late again.

5 **Put in the missing connectors from the box.**

although	because	in fact	so that	to begin with
and	furthermore	or	that	which

A lot of young singles are sharing flats in Dublin ... [1] rents are so high there. ... [2] some people don't like the idea of sharing, I think it it has many advantages. ... [3], costs are much lower ... [4] you can live in a more comfortable flat ... [5] in a better part of town. ... [6], sharing doesn't necessarily mean ... [7] you have no privacy at all. ... [8], each person has a room ... [9] is completely private ... [10] he/she only shares the bathroom, kitchen and living room.

A TALKING ABOUT JOBS

1

Warm-up

Link the photos of people at work 1–10 to the jobs in the box A–J.

2

3

4

5

6

7

8

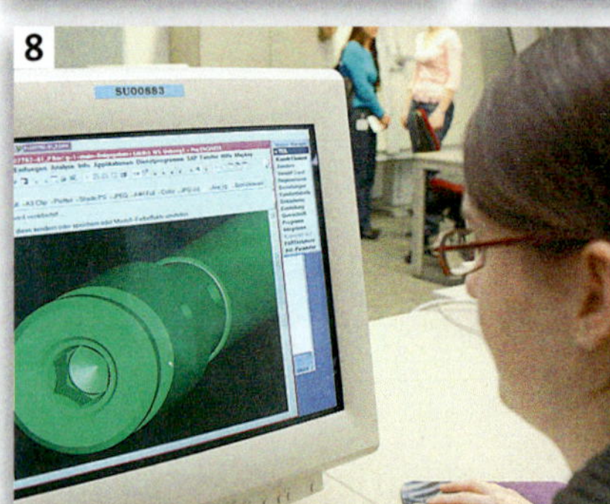

9

10

A carpenter	**E** electrician	**I** truck driver
B CNC lathe operator	**F** heating technician	**J** welder
C computer technician	**G** lab assistant	
D draughtsperson	**H** mechanic	

1 Look at the ten jobs. Would you be interested in any of them? Which? Why?

Which job	do you find	most least	interesting? worthwhile?	
	would you	be most interested in doing yourself?		
		like to	do have	most? least?
Do you find any of the jobs in the list interesting?				
Why do you say that?				
Why don't you find any of the jobs		interesting? worthwhile?		
I'd be interested in	becoming working as	a computer technician an electrician ...	because it's	future-orientated. interesting. varied. secure. well-paid. ...
I'd like to	work as			

2 Look at the tools and equipment and use the material to name them and say what they are for.

1

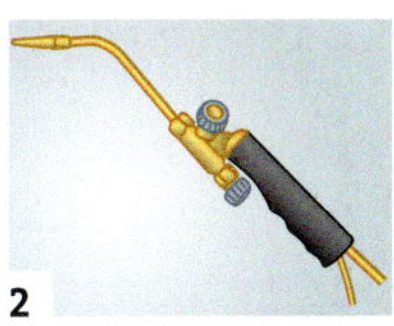
2

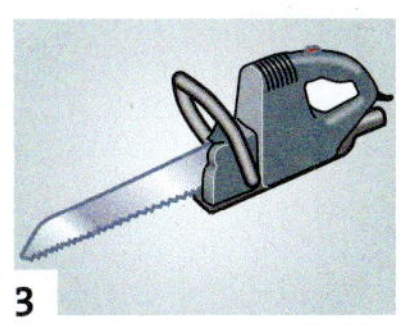
3

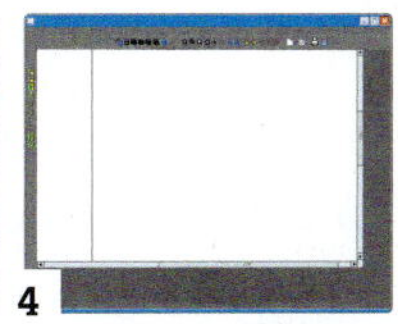
4

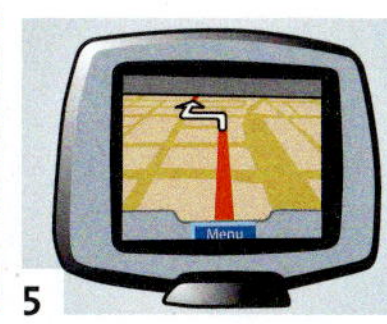
5

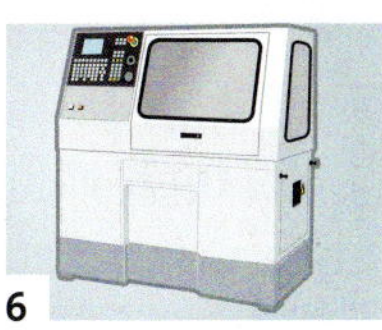
6

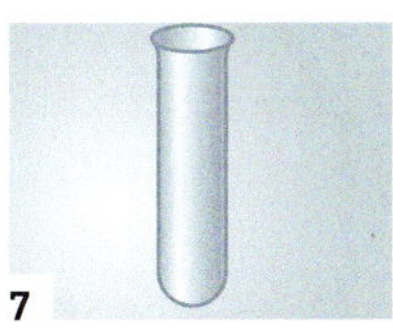
7

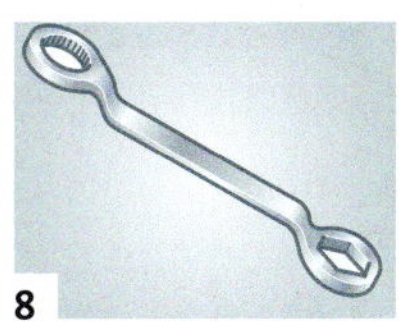
8

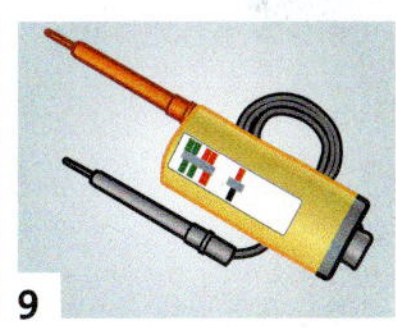
9

10

What	does	the	first second ...	drawing illustration picture	show?

The	first ...	drawing ...	shows is of	a	CAD screen. / hard disk drive. / CNC lathe. / pipe wrench. / power saw. / ring spanner. / satnav. / test tube. / voltage tester. / welding torch.

This	tool equipment	is used to	tighten or loosen hexagonal nuts and bolts. / cut materials into a round or cylindrical shape. / design products using computer software. / store and access digital data. / measure the voltage of a circuit. / join metals, e.g. pipelines, with welded joints. / hold and heat small amounts of chemicals. / show a route using a GPS system. / cut a hard material like wood or metal. / join or separate pipes with threaded collars.
A	CAD screen ...		

www.jobdescription.com

Are you interested in a technical occupation?
Click an occupation in the menu to call up other job descriptions.

Job description

Electrician

A

B

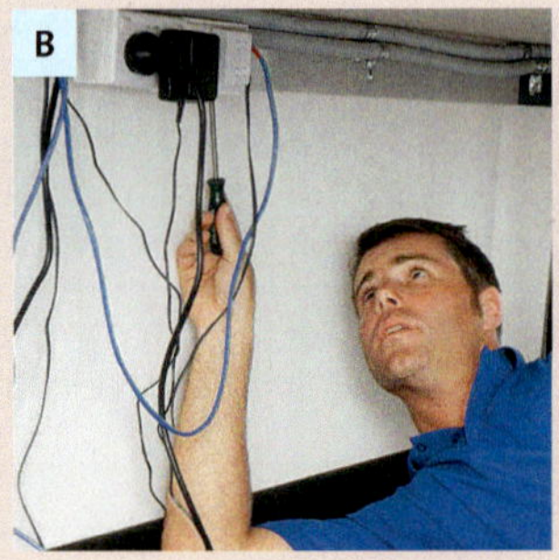

C

Electricians are involved in the generation, transmission and use of electricity.

Duties

Electricians install, maintain and repair electrically-operated equipment such as generators, electrically-powered machines, water and space heaters, and a huge variety of electric appliances from washing machines to toasters.
In the construction industry, electricians must ensure that the electricity supply and wiring are correctly installed. By law, only qualified electricians are allowed to connect power cables to electric motors and to high-voltage appliances such as cookers, and to service such equipment.
In the generation and distribution of electricity, electricians are responsible for the erection and maintenance of pylons, the connection of high-voltage power lines and the installation and maintenance of the transformer substations that reduce voltage to, in Europe, 230 volts.

Requirements

In addition to having the required technical qualifications, electricians must be careful, responsible and thorough as their work can be dangerous. They must be able to remain calm and accurate when working under pressure and never take short cuts to save time. They must be physically fit and have good eyesight and hearing.

Employment prospects

We live in an "electric world" so there is always a big and growing demand for electricians in the manufacturing and construction industries, as well as in transport and communications.

Qualifications | Courses | Job vacancies

3 **Answer the questions.**

1 Why are electricians so important in the construction industry?
2 How do we know from the job description that high-voltage electricity is particularly dangerous?
3 Why must electricians be careful and thorough?
4 Why do electricians obviously need good eyesight?
5 Why are electricians' job prospects so good?
6 Could you imagine becoming an electrician? Say why/why not.

4 **a) Find the noun forms of the verbs in the job description.**

1 (to) describe
2 (to) distribute
3 (to) employ
4 (to) equip
5 (to) generate (2 forms)
6 (to) heat
7 (to) maintain
8 (to) qualify
9 (to) transform
10 (to) transmit

b) Complete the sentences with the correct noun.

1 Coal-powered electricity ... plants are one cause of climate change.
2 You need a 12-volt ... to charge mobile phones.
3 People spend a lot on *buying* cars but not much on their ...
4 In the USA, paper ... are often less important than personality and experience.
5 In electricity plants, the ... are powered by steam.
6 Aren't distribution and ... almost the same thing?

5 **Writing**

Write a job description of a technical occupation and then present it to the class. Use the same headings as in the text. This website could help you: www.jobdescription.com.

6 Tr18 **Listen to four people talking about their part-time jobs. Take notes while listening and then fill in the table.**

Name	Age	Workplace	Hours/Week	Pay per hour + extras
Nick King	a) ***	b) ***	c) ***	d) ***, no extras
e) ***	25	f) ***	g) ***	h) ***, no extras
i) ***	22	electronics store	j) ***	k) *** + l) £*** Saturday bonus
Avi Gupta	m) ***	n) ***	o) ***	p) *** + ***% staff discount

7 **Do you have or would you like a part-time job? Why/Why not?**

Personality test

Many American employers use this personality test to get a better picture of their job applicants. Do the questionnaire to find out what kind of person **you** are.

What to do

1 Answer each question **without** 'reading ahead'. Answer **all** the questions.
2 Be honest. Answer as the 'real you'.
3 When you have finished, add up your score and see what it means.

Questions

1 You feel at your best
A in the morning.
B during the afternoon and early evening.
C late at night.

2 You usually walk
A quite fast with long steps.
B quite fast with short steps.
C less fast with head up.
D less fast with head down.
E very slowly.

3 When talking to people, you
A stand with your arms folded.
B have your hands clasped.
C have one or both your hands on your hips.
D sometimes touch the other person.
E play with your ear, touch your chin or smooth your hair.

4 When relaxing, you sit with
A your knees and legs neatly together.
B your legs crossed.
C your legs stretched out.
D one leg curled beneath your body.

5 When you find something funny, you
A open your mouth wide and laugh loudly.
B open your mouth a bit and laugh less loudly.
C chuckle quietly.
D just smile.

6 When you go to a party, you make a

A noisy entrance so everybody knows you've arrived.
B quiet entrance, looking around for somebody you know.
C very quiet entrance, hoping that nobody notices you.

7 You're concentrating on something and somebody interrupts you. Do you

A welcome the break?
B feel very annoyed?
C move between these two extremes according to who it is?

8 In this list, the colour you like best is

A red or orange.
B black.
C yellow or light blue.
D green.
E dark blue or purple.
F white.
G brown or grey.

9 In the last few moments before you go to sleep, you lie

A stretched out on your back.
B stretched out on your stomach.
C on your side with knees slightly bent.
D with your head on one arm.
E with your head under the covers.

10 When you are asleep, do you

A often dream that you are falling?
B often dream that you are fighting against something?
C often dream that you are looking for something or somebody?
D often dream that you are flying or floating in the air?
E usually have a dreamless sleep?
F always have sweet dreams?

Scoring

	A	B	C	D	E	F	G
1	2	4	6	-	-	-	-
2	6	4	7	2	1	-	-
3	4	2	5	7	6	-	-
4	4	6	2	1	-	-	-
5	6	4	3	5	-	-	-
6	6	4	2	-	-	-	-
7	6	2	4	-	-	-	-
8	6	7	5	4	3	2	1
9	7	6	4	2	1	-	-
10	4	2	3	5	6	1	-

What your personal score says about you

Over 60 People see you as rather dominant. You like to take control and order others about. Some less self-confident people may wish they could be a little more like you sometimes, but they don't really want to have too much to do with you.

51–60 Other people see you as an exciting risk-taker. People may find life more colourful when you're around, but they also think that you can be too impulsive and reckless. They certainly don't have much confidence in your judgement.

41–50 Others see you as fresh, lively, charming, amusing, practical and never boring. You like to be the centre of attention, but you do not let your popularity go to your head. People also find you kind and understanding.

31–40 Others see you as sensible, careful and practical. If you are clever or talented, you are modest about it. You do not make friends easily but you are extremely loyal to your small circle of close friends – and you expect the same loyalty from them.

21–30 Other people see you as a painstaking, cautious plodder. They expect you to examine everything very carefully and then usually decide against it. This is because your careful nature always makes you play safe.

Below 21 People think you are shy and afraid of making mistakes or upsetting people. You always want someone else to make the decisions. Some people think you're too quiet and a worrier, but those close to you know that you have hidden depths that you seldom show.

8 **There are six categories of people in the scoring of the personality test above. Fit the jobs in the list to the categories. Give reasons for your choice.**

electrician	photographer	security guard
handball coach	plumber	software engineer
lab assistant	police officer	soldier
museum assistant	racing driver	teacher

B FOCUS ON TECHNOLOGY: MATERIALS

Materials

A **material** is anything that we use to make a product. We generally use glass or plastic to make bottles, for example.

9 **Work in small groups. Brainstorm products that are often made of the materials in the box. Then compare your list with that of other groups.**

aluminium, cardboard, ceramic, concrete, cotton, fibreglass, foam rubber, glass, gold, leather, paper, plastic, rubber, steel, stone, tin, wax, wood, wool

10 **Complete the text with words or expressions from the list.**

composite material	paint	synthetic material
concrete	plastic	wood
glass	processed material	
natural material	steel	

In the three photos, the carpenter is working with ... [1]. The lathe operator is turning a piece of ... [2]. The assembly line worker is fitting ... [3] components to a car.

The three people are all using very different materials. One of the materials, wood, is a ... [4] that we find in nature and can use more or less immediately. Another, steel, comes from a natural or 'raw' material but we must process it before we can use it. It is a ... [5]. The third material does not occur in nature at all – it is a man-made or ... [6].

In practice, there are very few natural materials that we can use without processing them first. Often, we mix several kinds of materials to make a mixed or ... [7] such as ... [8], ... [9] and ... [10], for example.

11 Work with a partner. Look at the products below. Which are made from natural materials? Which are made from processed or synthetic materials?

apples

polyester paint

beer glass

paper bag

brass screw

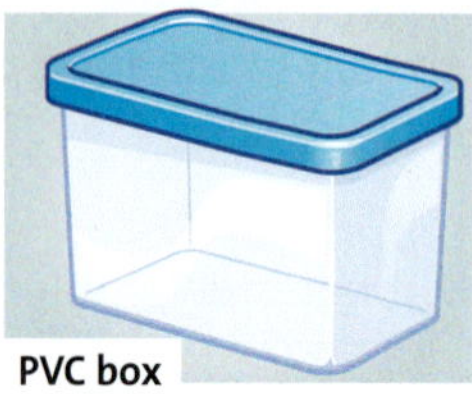
PVC box

tyre

petrol can

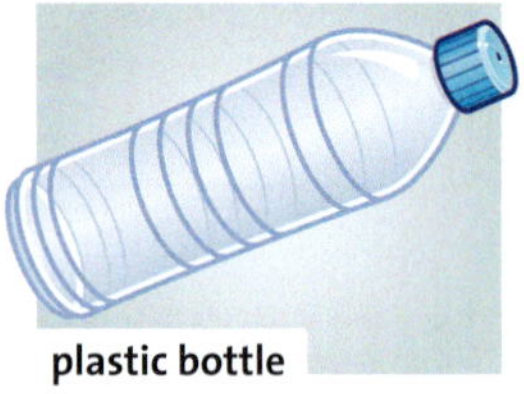
plastic bottle

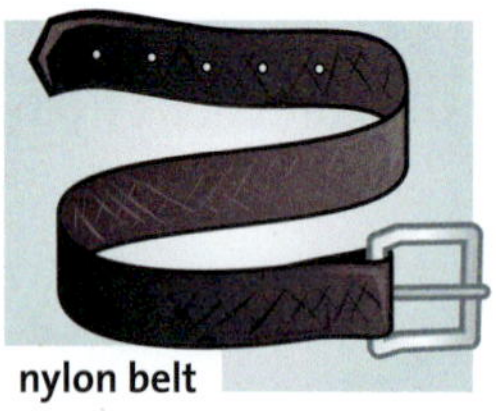
nylon belt

cotton T-shirt

aluminium piston

12 **a)** We use these adjectives to describe the properties of materials. Link the adjectives 1–8 to their meanings A–H and give their German equivalents.

1 brittle
2 ductile
3 durable
4 hard
5 impact-resistant
6 light
7 non-corrosive
8 transparent

A bends easily without breaking
B can be seen through
C does not break or crack easily when hit
D does not react with chemicals easily
E hard, but cracks/breaks easily when hit
F lasts a long time before it wears out
G opposite of 'heavy'
H opposite of 'soft'

Tr 19 **b)** What are the pros and cons of ceramic hip implants compared with plastic or metal ones? First, copy the grid below into your exercise book. Then listen to the radio talk and complete it.

Material	Pros	Cons
plastic		
stainless steel, nickel, titanium		
ceramic		

Joining materials together

In manufacturing and construction, materials often have to be joined together. There are several ways of doing this, but the most common are **pinning** and **bonding**.

Pinning is done with a **fixing device** such as a nail, a screw, a rivet, a staple, a bolt or a clip. Bonding is done with a **chemical fixing agent** such as an adhesive or by applying pressure or heat, as in welding.

When we want to make a very strong or 'heavy-duty' joint, we often use more than one joining method. A furniture-maker, for example, often joins the legs of a chair to the seat by bonding them with a wood glue and then pinning them with pegs or brass screws.

13 Look at the pictures. Complete the sentences with words from the box.

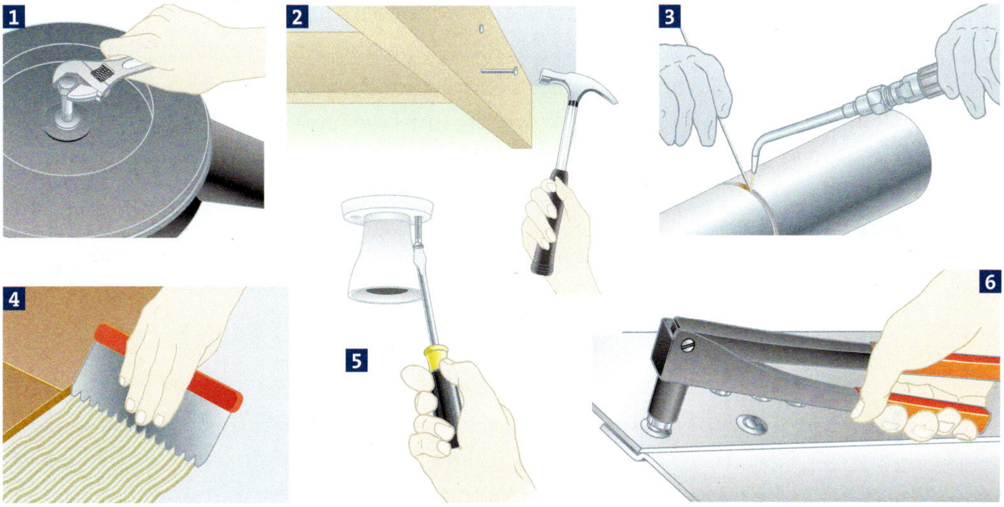

1 The mechanic is using a ... to fix the air filter together.
2 The carpenter is fixing the beams together with ...
3 The welder is using a ... to join sections of a pipe.
4 The tiler is laying floor tiles with a special ...
5 The electrician is fitting the lamp-holder to the ceiling with ...
6 The riveter is using ... to fix steel plates together.

adhesive
bolt
nails
rivets
screws
welding torch

Practising grammar

Remember:
Some, *any* and their compounds → Begleitgrammatik, Abschnitt 9, Seite 154

1 **Some** materials are completely synthetic.
2 You often like **somebody** to tell you what to do.
3 Do you have **any** information about modern plastics?
4 I don't know **anything** about the properties of titanium.

1 Complete the sentences with *some* or *any*.

1 ... metals are too soft for use in implants.
2 Does Peter Becker have ... industrial experience?
3 ... materials are made by mixing several other materials together.
4 Did you visit ... flats at the weekend, Anton?
5 I found ... interesting information about hydroelectric plants in Austria.
6 We haven't had ... applications for the job yet.
7 We can't service ...[1] forklifts until we have ...[2] special oil.
8 A security guard saw ...[1] men taking photos of ...[2] new machines.

2 Complete the sentences with *some*, *any* or one of their compounds.

1 **A** Listen! I think I heard ... go into the office.
B Really? I didn't hear ...
2 **A** I can't find ... drill bits.Do you have ...?
B Sorry. I don't have ..., either. Ask Markus. Perhaps he has ...
3 **A** Have you seen Tara ...? I have ... messages for her.
B Sure. She's ... in the factory with ... visitors from Vienna.
4 **A** Do we have ... to drink?
B Well, we have ... mineral water but I expect you'd like ... stronger.
5 **A** I need ...[1] to help me repair this door. Can you think of ...[2] I could ask?
B Ask Tom. And we need ...[1] more paint. If he isn't doing ... [2], he can get some.

Remember:
Countable and uncountable nouns → Begleitgrammatik, Abschnitt 10, Seite 154

1 **Materials** are divided into a number of **categories**.
2 **Rubber** and **wood** are two examples of natural materials.
3 What **kinds of rubber** are suitable for car tyres?
4 There's **a can of oil** on the workbench.

3 **In your exercise book, draw a table for countable and uncountable nouns like the one below. Put the nouns in box A into the correct column (i.e. countable or uncountable).**

Now give the plural form of the countable nouns and add a suitable countable expression from box B to the uncountables so that you can use them in the plural.

A

box	equipment	glass	oil	rubber
bus	evidence	glue	paint	water
car	experience	knife	people	wood
dynamo	furniture	news	pliers	wrench

B

pieces of	cans of	pairs of
bottles of	crowds of	tubes of

Countable		Uncountable	
Singular	**Plural**	**Base form**	**Qualified form**
box	*boxes*	*equipment*	*pieces of equipment*
...	...	...	...
...	...	...	...
...	...	...	...
...			...

Remember: *Much, many, a lot of* → Begleitgrammatik, Abschnitt 11, Seite 155

1 In practice, there are not **many natural materials**.
2 How **many synthetic materials** can you think of?
3 Plastics are not **much use** in very hot conditions.
4 Does the internet use **much electricity**?
5 We need a material that can stand **a lot of heat**.

4 **Use *much, many* or *a lot of*.**

1 Hurry up! We don't have ... time.
2 How ... people are there in your class?
3 You won't find ... materials suitable for this application.
4 CSS employs quite ... women but not ... of them are managers.
5 A: How ... are we spending on wages at the moment?
B: Too ... We are going to lay off (*entlassen*) ... workers, I'm afraid.
6 Tom has ... experience. You can learn ... things from him.
7 Modern cars don't need ... maintenance.
8 There were ... skiers on the slopes today.

A ARRANGING AND TAKING PART IN MEETINGS

Warm-up

Look at the cartoons and speech bubbles. Then answer the questions.

1. What's the general message?
2. What do you think the result of the boss's questions was?
3. Two of the items are true. Which ones?
4. Have you ever taken part in a meeting? Did you find it useful?

A checklist

Tony Harper is the production manager at a company that makes parts for the car industry. There have been a growing number of quality complaints and now Tony wants to arrange a meeting to review the company's QA systems. First, he asks his assistant, Lynn Miller, to make a checklist.

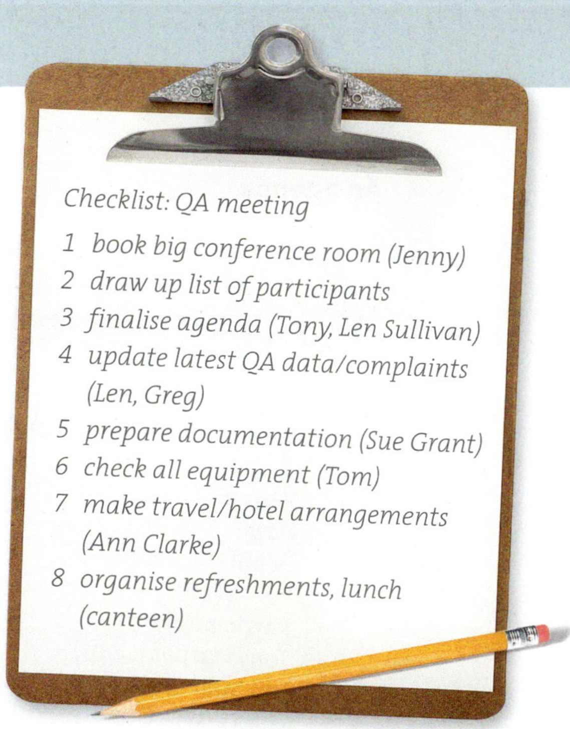

1 Work with a partner. One of you is Lynn, the other one is a person from the checklist. Sit back to back and use the German prompts to make short telephone calls. Swap roles once.

1. Lynn ruft **Tom** an. Sie bittet ihn, alle Geräte für die Sitzung aufzustellen und auf ihre Funktionsfähigkeit zu überprüfen.
   Tom antwortet, dass er schon alles überprüft hat und die Geräte in Ordnung sind. Lynn braucht sich keine Sorgen zu machen.
2. Lynn ruft **Anne Clarke** an. Sie fragt, ob sie ihr bei den Hotelreservierungen helfen kann. Sie benötigt zwei Einzelzimmer für Jens Herzog von VW und Luc Picard von Citroën für eine Nacht.
   Anne antwortet, dass sie gerne helfen würde, aber sie hat erst heute Nachmittag Zeit.
   Lynn dankt ihr und sagt, dass das in Ordnung geht.
3. Lynn ruft **Len** an, um zu fragen, ob die QA-Daten auf den neuesten Stand gebracht worden sind und rechtzeitig für die Sitzung fertig sein werden.
   Len antwortet, dass sein Mitarbeiter Greg dabei ist, dies zu tun und alles fertig sein wird.
   Lynn dankt ihm.
4. Lynn ruft in der Kantine an und spricht mit **Mrs Brown**. Sie sagt ihr, dass sie die üblichen Erfrischungen um 10 Uhr 30 und um 15 Uhr im großen Konferenzzimmer im 3. Stock benötigen werden.
   Mrs Brown sagt, dass dies in Ordnung geht, und fragt, ob sie nur Kaffee oder auch Tee möchten.
   Lynn antwortet, dass auch Tee gut wäre, and fragt nach dem Mittagessen.
   Mrs Brown erwidert, dass das Mittagessen mit einer Auswahl von drei Gerichten schon bestellt ist.

An agenda

Pegasus Automotive Limited

Meeting to evaluate QA systems

Venue: Big conference room, 3rd floor
Date: 28 October 20..
Begin: 9.00 a.m.

Participants
Tony Harper (chair)
Len Sullivan, Quality Assurance Manager
Greg Davis, Quality Assurance Technician
Dick King, Quality Assurance Technician
Mary Johnson, Customer Services Manager
Alice Brown, Dispatch & Logistics
Lynn Miller (minutes)

Jens Herzog, Volkswagen (morning only)
Luc Picard, Citroën (morning only)

Agenda

Item	Topic
1	Greetings, introductions
2	Programme for the day
3	Aim of meeting
4	Review of current situation
5	Discussion of current QA systems
6	Identification of problem areas
7	Future action
8	AOB*

Tony Harper

Tony Harper
Quality Assurance Manager

* Any Other Business

2 **The eight sentences are all from the meeting. Link them to a) the speaker and b) the correct item on the agenda. Look at the example first.**

1 "Jens Herzog and Luc Picard are here to give our customers' point of view. Thank you very much for coming." *Tony, agenda item 1*
2 "Any other points? No? Then I'd like to thank you all for coming. Lynn will email the minutes to you within the next few days."
3 "By the end of this meeting we hope to have some answers to our current quality problems."
4 "I'm no expert on QA, but we have begun to solve our delivery problems in three ways. Let me describe them."
5 "Okay, Tony, but who's responsible for doing all this? QA or my department? I can see problems here."
6 "Tony has asked me to tell you about the arrangements for today. Well, we will have a 15-minute break at about 10.30 and then ..."
7 "Volkswagen had a similar problem in Zwickau and we solved it with improved training."
8 "Well, Tony, the latest complete QA data we have is for the last quarter. Greg, could you say a few words, please?"

3 **How would you express the German words or expressions in English?**

1 There were six (*Teilnehmer/innen*) during the afternoon session.
2 Tony asked Lynn Miller to write the (*Protokoll*).
3 Tony called the meeting because of (*Qualitätssicherungsprobleme*).
4 The (*Ziel*) 1 of the meeting was to (*bewerten*) 2 Pegasus's QA systems.
5 (*Tagesordnungspunkt*) 1 four was about the (*aktuelle Lage*) 2.
6 After discussing the (*Problemzonen*) 1, the meeting talked about (*zukünftige Maßnahmen*) 2.

4 **You are Lynn Miller. Listen to the extract from the meeting.**
Tr 20 **Take notes for the minutes under these headings:**

1 Problem areas
2 Action to improve situation
3 Names of people responsible for action
4 Date by which action should be complete

Now write Lynn's minutes from your notes.

B SHAPES

There are two main types of shapes: **plane shapes** and **solid shapes**.

Plane shapes have two dimensions: **height** and **length**:

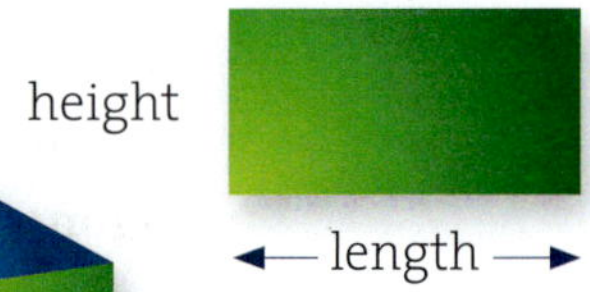

Solid shapes have three dimensions: **height**, **length** and **width**:

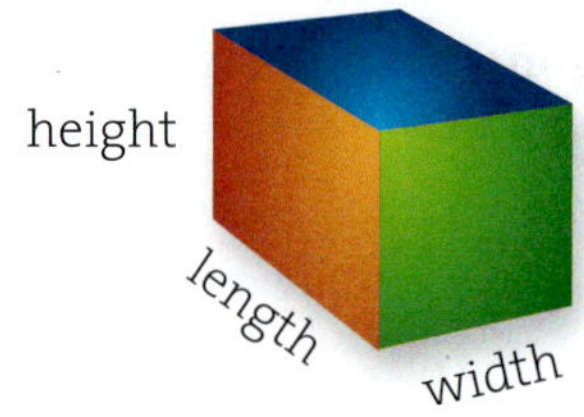

Both plane and solid shapes fall into two further groups: **rectilinear shapes** and **curvilinear shapes**.

Rectilinear shapes have **straight sides and angles**. Curvilinear shapes have **curved sides** and **no angles**.

Length of sides and angles

If we give no exact dimensions, then we mark the length of sides with **a**, **b**, **c**, **d**, ... Sides with the same letter have the same length or 'value'.

In the same way, we mark angles with Greek letters α, β, γ, δ, ...

Look around the room. You will see that in practice most shapes are made up of a combination of straight and curved lines. These are called **mixed shapes**.

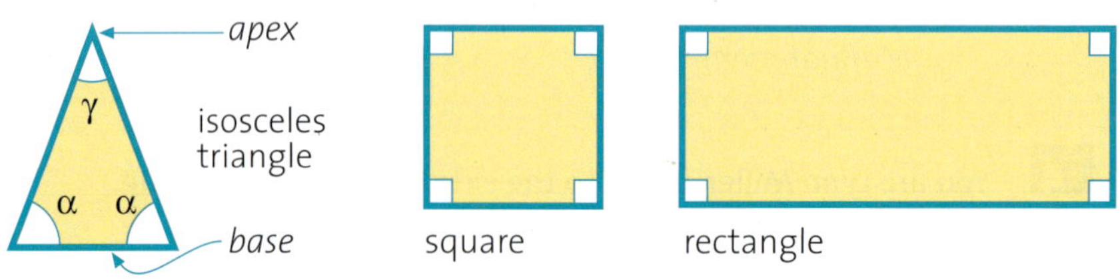

All rectilinear shapes have **angles**. The value of angles is expressed in **degrees** (°).

A **triangle** has three angles, two at the **base** and one at the **apex**. The **combined value** of the three angles is always 180°.

Four-sided shapes have four angles, one at each corner. The combined value of the four angles is always 360°.

Squares have four sides of equal length and four angles of 90°. **Rectangles** also have four 90° angles, but only the opposite sides are equal in length. The opposite sides of both squares and rectangles are **parallel** to each other.

Rectilinear shapes with more than four angles or sides are called **polygons**. The most common polygons are **pentagons** (five angles/sides), **hexagons** (six angles/sides) and **octagons** (eight angles/sides).

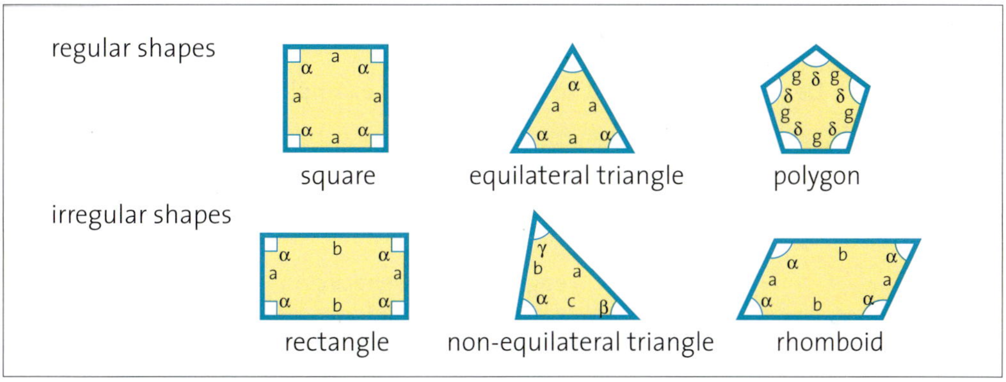

We also speak of **regular shapes** and **irregular shapes**.

Regular shapes have **sides of equal length** and **angles of equal value**. Squares, equilateral triangles and polygons are regular.

The sides and/or angles of irregular shapes are **not all equal in value**. Rectangles, all triangles other than equilateral triangles and rhomboids are examples of irregular shapes.

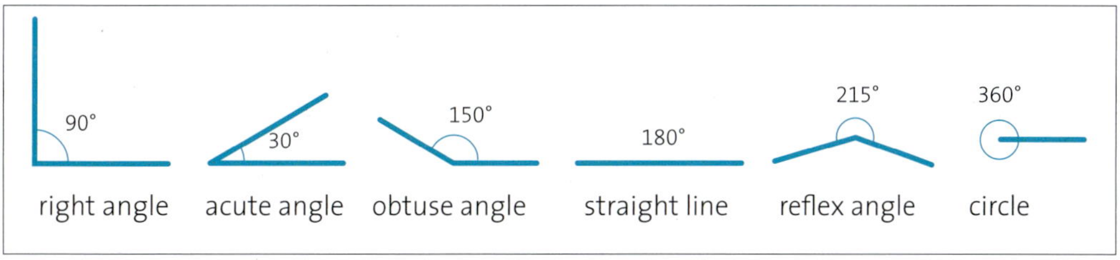

There are four different types of angle. A **right angle** is 90°. Squares and rectangles have four right angles.

Angles of less than 90° are **acute angles**. Triangles have either one right angle and two acute angles, one obtuse angle and two acute angles or three acute angles.

An **obtuse angle** has a value of more than 90° but less than 180°. (An angle of 180° is a **straight line**.) Rhomboids always have two acute angles and two obtuse angles.

We call angles with a value of more than 180° but less than 360° **reflex angles**. (An angle of 360° is a **circle**.)

Giving dimensions

We give the dimensions or measurements (*Abmessungen*) of rectilinear solid shapes in the reverse alphabetical order – **width** (side to side), **height** (top to bottom), **depth** (front to back), for example:

41.5 cm × 14.75 cm × 28 cm

We say: forty-one point five **by** fourteen point seven five **by** twenty-eight centimetres.

5 **Describe the shapes 1 to 12. Look at the examples first.**

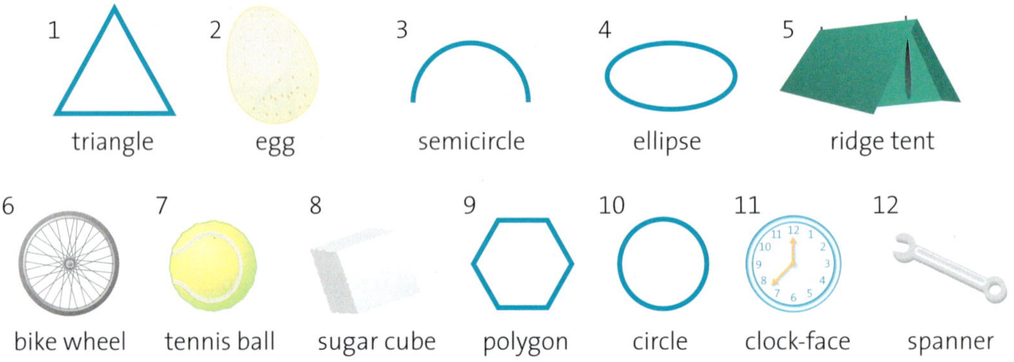

- The *first* picture is of *a triangle*. It is a *rectilinear plane* shape.
- The *second* illustration is of *an egg*. It is a *curvilinear solid* shape.
- The *third* ...

6 **Find words and expressions in the text that fit the definitions. They are not necessarily in the same order.**

1 an angle of 180°
2 an angle of 90°
3 any shape with straight sides
4 plane shape with four equal sides
5 shape with more than four angles
6 shapes whose sides and angles are not all equal
7 top of a triangle
8 three-dimensional shape
9 Two lines that are the same distance apart along their whole length are ...
10 an angle that is greater than 180° and less than 360°
11 a triangle with three equal angles
12 an eight-sided regular shape

7 **Work with a partner. Fill in the missing prepositions. Swap roles once.**

1 A triangle is made ... ... three sides.
2 Shapes ... four sides have an angle ... each ... the four corners.
3 Right angles are angles ... 90°.
4 The opposite sides of a rhomboid are parallel ... each other.
5 The four angles of a square add up ... 360°.
6 The value ... angles is expressed ... degrees.
7 This box is 30 centimetres ... seventy ... twenty.
8 Triangles have an angle ... the apex and two ... the base.

8 **Look at the close-ups. What are they?**

pencil	screwdriver
saw	slimline TV
screw	USB stick

1

2

3

4

5

6
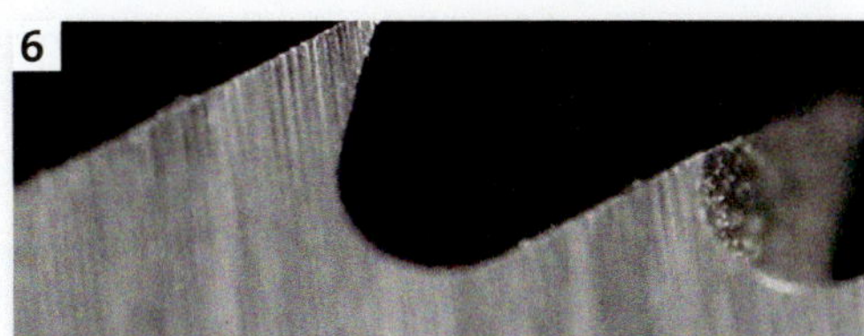

9 Tr 21

a) Listen to the description of a common device. Then listen again and draw it. What do you think it is?

b) Work with a partner. Describe the shape(s) of any common tool to your partner – for example a socket wrench or a saw. Your partner should draw the shape(s) you describe and name the tool. Then reverse roles.

Practising grammar

Remember: Relative pronouns → Begleitgrammatik, Abschnitt 12, Seite 155

1 Lynn was the secretary **who/that** wrote the minutes.
2 Pegasus is a British company **which/that** supplies parts to the car industry.
3 Björn is a German technician **whose** uncle lives in Leipzig.
4 Is this the computer **whose** battery needs changing?

1 **Fill in *who, which* or *whose*.**

1 Harry Winter is the man ... is responsible for this project.
2 The LX100 is the new machine ... broke down yesterday.
3 The man ... car I damaged is my boss!
4 The guy ... tried to ring you is going to ring again this afternoon.
5 The firm ... won the contract to build the new stadium is Austrian.
6 A woman ... name I forgot to ask wanted to speak to you.

Remember: Modal verbs → Begleitgrammatik, Abschnitt 13, Seite 156

1 Electricians **must be** careful, responsible and thorough.
2 We **mustn't forget** to tell Len about the meeting.
3 You **needn't come** to the meeting, Harry. Greg's coming.
4 Lynn **can you write** the minutes, please?
5 It isn't certain yet, but Pegasus **may open** a plant in Romania next year.
6 Spot checks aren't enough. We **should inspect** everything that leaves the plant.

2 **Use *must, mustn't* or *needn't*.**

1 We ... (send) any faulty goods to our customers.
2 You ... (write) the report, Greg. I'll do it.
3 I ... (work) all day Sunday so I ... (come) to work on Monday.
4 You ... (forget) to book two hotel rooms, Lynn.
5 I'm afraid we ... (cancel) the meeting if Tony can't come to it.
6 Len ... (go) to Graz tomorrow. The problem has been solved.

3 **Think of your lifestyle. Say what you *must/mustn't* do, what you *needn't* do and what you *could, should* or *may* do. The words in the box will help you.**

cycle/walk to school/the shops/work/...	take the dog for a walk
do more exercise/lose weight/take up a sport	(not) use buses/lifts/stairs/the car
(not) drink alcohol/mineral water/sweet drinks	(not) visit so many chat rooms
(not) eat fast food/healthy food/sweets	(not) watch so much TV
(not) spend so much time surfing the internet	work harder

4 **You are a member of a club which is planning a three-day motorcycle tour with two nights at a campsite. Tonight you are going to a meeting to plan the tour. Use the material to prepare for the meeting.**

I	could	buy a new road map.
We	must	check the first-aid kit.
Somebody	should	choose nice places for a break.
	must remember to	decide on the best route.
	mustn't forget to	make a route map.
		find a good campsite.
		find out about places of interest on our route.
		fix a time and place to start the tour.
		photocopy the route map and programme.
		reserve places at the campsite.
		tell everybody what to bring with them.
		try to avoid roadworks on the route.

Start like this: *We must decide on the best route. And somebody should make ...*

Remember: Future with *going to* → Begleitgrammatik, Abschnitt 14, Seite 156

1 We **are going to open** a plant in Romania next year.
2 Don't worry. They **are not going to close** the plant in Dublin.
3 **Is** Len or Greg **going to write** the report?
4 Look at those clouds. It**'s going to rain** in a minute.

5 **Use the correct form of the *going to*-future – positive, negative, questions.**

1 Tony ... (arrange) another meeting in a month.
2 ... (that computer/overheat) without a fan?
3 Watch out! Those pallets ... (fall) over.
4 ... (Tony/ask) Lynn to write the minutes again?
5 It isn't fair. I ... (not write) the minutes again, Tony.
6 We can go cycling today. The forecast says it ... (not rain) after all.
7 The trainees won't be here tomorrow. They ... (visit) a trade fair.

6 **Translate the sentences into English.**

1 Sie können dieses Navigationsgerät nur in Europa benutzen.
2 Wir brauchen am Samstag nicht zu arbeiten.
3 Gregor, kannst du auch zu der Sitzung kommen?
4 Darf ich Ihnen/dir/euch Laura Oberleitner aus Österreich vorstellen?
5 Du solltest die Fabrik nie ohne Schutzhelm (*hard hat*) betreten.
6 Zuerst müssen wir kontrollieren, ob die Maschine betriebsbereit (*operational*) ist.

Warm-up

Do the assignments.

Tr 22

1 Listen to the sounds you hear in a typical household. Use the numbers 1–10 to link the sounds to a product from the box. Note that two products don't fit at all.

bottom-hung window	dishwasher	toaster
coffee maker	door	toilet and cistern
combination boiler	outside blinds	washbasin with plughole
computer	shower	washing machine

2 Sort the products into the correct category, **a**) household appliances and fittings, **b**) plumbing/heating equipment.

3 What are the products used for?

4 Which products do you often/sometimes/never use? Do a class survey.

Link the drawings 1–7 to the explanations A–G.

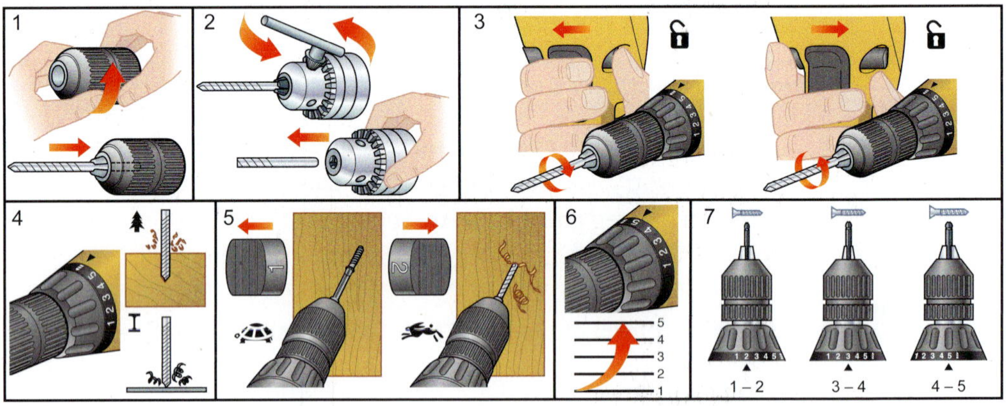

A To remove a drill bit from a keyed chuck, fit the key to chuck as shown and turn it in an anticlockwise direction to open jaws.

B To reverse the sense of rotation, press the locking button from right to left (cw=clockwise travel) or left to right (ccw=counterclockwise, anticlockwise travel).

C To change the speed of the drill, push the slider switch from 1 (slow) to 2 (fast) and vice versa (*umgekehrt*).

D To use the machine in screwdriver mode, turn the setting ring to the **numbered** section.

E To adjust the torque according to the size of the screwdriver bit/screw, turn the setting ring to the appropriate setting (number), as shown in the drawing.

F To insert a drill bit in a keyless chuck, turn the outer collar of the chuck in an anticlockwise direction to open jaws.

G To use the machine in drill mode, turn the setting ring to the end of its travel in a clockwise direction so that the **unnumbered** marker is opposite the arrowhead.

A Heat pumps

Unlike a gas-fired boiler, for example, a heat pump does not burn any fuel to produce heat,, and the amount of electricity it consumes is low compared with the consumption of an electric heating system. This characteristic makes heat pumps exceptionally friendly to the environment.

Technically, heat pumps function on the same principle as refrigerators and air conditioners. A liquid absorbs heat as it turns into a gas, and releases heat as it returns to its liquid state.

This is how this cycle works in a heat pump:

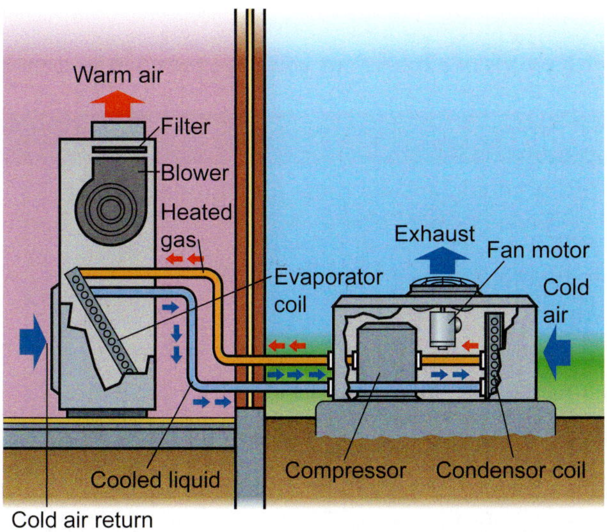

During the warm summer months, the heat pump operates as an air conditioner. It removes heat from a house and releases it outside. A liquid refrigerant is pumped through a coil of tubing called the 'evaporator'. The liquid expands as it moves through the coil. The liquid changes into its gaseous state as it absorbs heat from the air around the evaporator coil.

A fan then pushes air around the cooled coil and through ducts into the house. Meanwhile, the hot gaseous refrigerant moves through a compressor and begins to liquify. It then moves to a condensor coil on the outside of the house, where the compressed gas releases its heat into the atmosphere and returns to its liquid state. Then the cycle begins again.

During the cool winter months, the heat pump reverses this process. It extracts heat from the air outside and releases it inside the house.

Heat pumps are very efficient when the outside temperature is within the range 7°C to 10°C, but they become progressively less efficient as the temperature drops. In fact, when the outside temperature falls significantly below about 7°C, heat pumps have to be supplemented by an auxiliary heating system such as a solid-fuel or gas-powered stove.

For this reason, heat pumps are not practicable in regions with low winter temperatures such as northern Europe, but they are an ideal solution in southern Europe where ambient winter temperatures are higher.

1 **Answer the questions.**

1 Explain why heat pumps are environmentally friendly.
2 Point out the similarities between heat pumps, refrigerators and air conditioners.
3 Describe the heat pump cycle in winter.
4 Say why heat pumps are not a practicable heating solution in, say, Sweden.

2 **a) Find the English equivalents of the German words/expressions in the text.**

1 *Gasheizkessel* (ll. 1–5)
2 *Klimaanlage* (ll. 1–5)
3 *Kühlmittel* (ll. 11–15)
4 *Verdampferschlange* (ll. 11–15)
5 *Flüssigkeit* (ll. 11–15)
6 *gasförmiger Zustand* (ll. 16–20)
7 *Gebläse* (ll. 16–20)
8 *Luftrohr* (ll. 16–20)
9 *Kühlschlange* (ll. 21–25)
10 *Arbeitsablauf* (ll. 26–30)

b) Use six of the English words or expressions you have found in sentences of your own.

Product presentations

■ A ***product presentation*** should answer these ***four*** questions:

1 What is the name of the product or model?
2 What is the product for?
3 How does it work?
4 What are its advantages, e.g. design, cost, convenience, low maintenance, ease of operation, environmental benefits?

■ These ***five*** tips will help you to give an ***effective presentation***:

1 Never read from a script. Use a **prompt card** with key words and key points.
2 Look at your audience, but not always at the same person. Good speakers **scan the room** from side to side and back to front as they talk.
3 Don't stand like a statue or walk around the room. **Be relaxed** – imagine you are talking to a friend.
4 **Vary the volume** of your voice according to the points you want to get across. Mark important points by speaking more loudly and slowly, but don't shout.
5 Emphasise important points with **gestures**, but don't overdo it.

3 Tr 23 **The product presentation you are about to hear was given before an audience of the company's authorised dealers (*Vertragshändler*). Listen to the presentation once. Then listen again and take notes. Finally, write the speaker's prompt card.**

Word list

versatile *vielseitig*
require *verlangen*
outstanding feature *hervorstechendes Merkmal*
requirement *Anforderung*
occupier *Bewohner/in*
override *Eingriff (in automatische Steuerung)*

B Heating systems

Underfloor heating for comfortable heat

Due to the extremely large heated area, i.e. the whole floor of the room, underfloor heating works at a low surface temperature. This also ensures even heat radiation and temperature distribution. In underfloor heating systems, the essential heat is transmitted by heat radiation. This means that the air temperature in floor-heated rooms can be considerably lower than in rooms with conventional heating without any loss of comfort. And while humidity in radiator-heated rooms can be uncomfortably low (around 30 %), the atmospheric humidity in underfloor-heated rooms in winter is usually around 60 %. Moreover, because of the low air circulation in floor-heated rooms, flying dust and dust mites are not such a problem. This is a big advantage for people who are allergic to house dust.

Floor constructions and heating pipes

The ROTEX Monopex floor heating system makes use of system plates.

Advantages:

- Constant temperature due to exactly laid and guided heating tube
- Heating pipe is totally surrounded and covered by screed
- Excellent heat insulation to rooms or ground beneath floor
- Excellent sound insulation
- Easy to install
- Ecologically neutral as CFC*-free; fully recyclable

Monopex secco: Dry construction

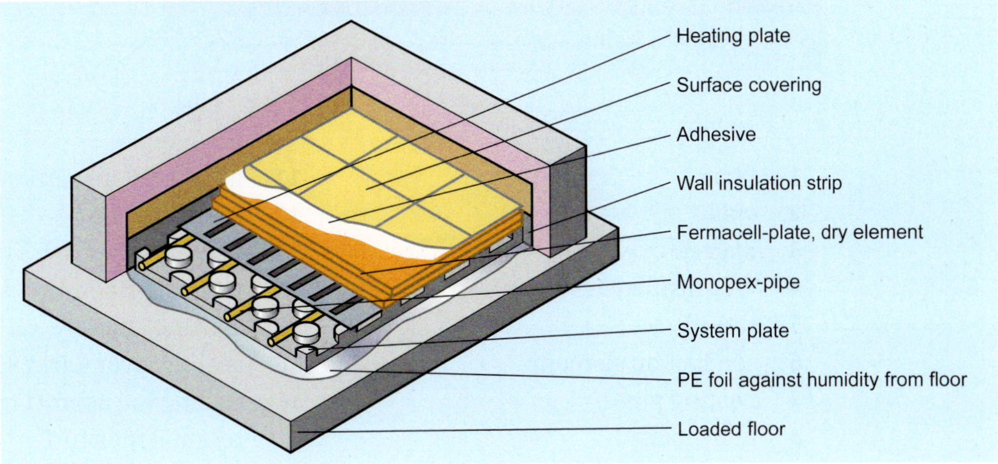

* **c**hloro**f**luoro**c**arbon *Fluorchlorkohlenwasserstoff* (FCKW)

Monopex secco: Dry system floor installations

Mono Plate: For floors above heated rooms

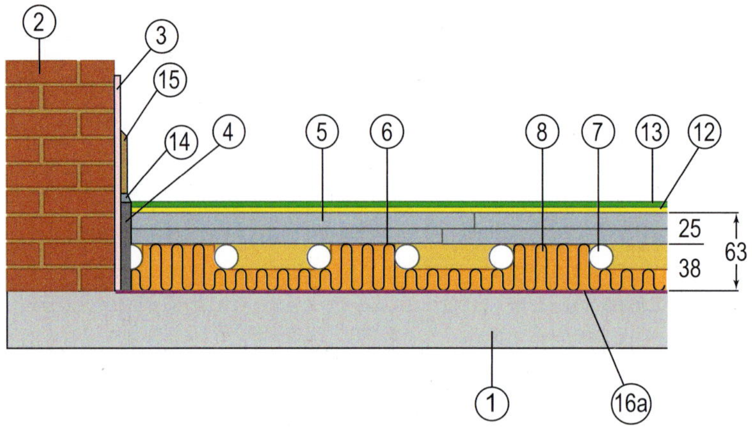

Compact Plate, 20 mm additional insulation: for floors above unheated rooms, soil, outside air

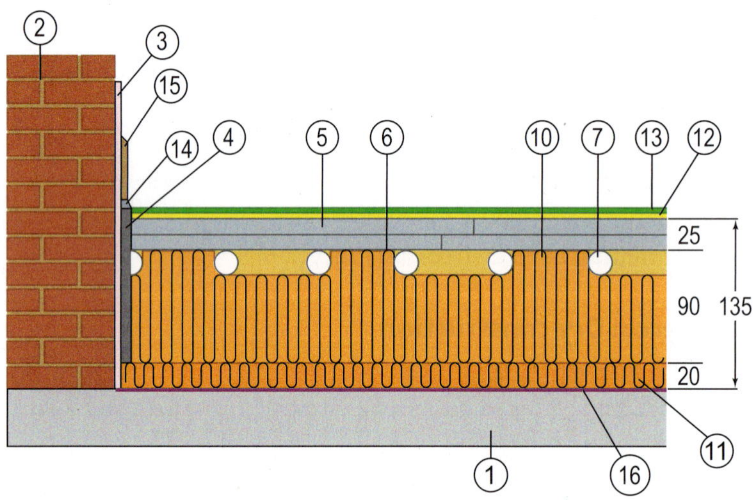

1	floor
2	wall
3	plaster
4	wall insulation strip
5	screed
6	conduction element
7	heating pipe
8	mono plate
10	compact plate
11	additional insulation plate
12	adhesive film
13	floor covering
14	elastic joint
15	skirting board
16	sealing layer, DIN 18195
16 a	sealing foil against humidity/ residual moisture

4 Answer the questions.

1 Say why this underfloor heating system works at a relatively low surface temperature.
2 Describe how heat is transmitted in underfloor heating systems.
3 Point out what effect a lower air temperature has on the comfort of people in the room.
4 Explain why people who are allergic to house dust may feel more comfortable with underfloor heating.

5 Complete the text with the missing prepositions.

The underfloor heating system operates ... [1] a relatively low surface temperature because the heated area is equal ... [2] the whole area ... [3] the floor. This also ensures an even distribution ... [4] heat ... [5] all parts ... [6] the room without the development ... [7] so-called "cold zones". Furthermore, ... [8] such heating systems heat is transmitted ... [9] heat radiation. This means that the air temperature can be lower without any loss ... [10] comfort ... [11] the people ... [12] the room.

6 Writing

Look at the drawing of a Monopex secco floor and describe its structure from the top down.

The surface floor covering – floor tiles or a fitted carpet, say – is fixed to the Fermacell plate with a layer of adhesive. Immediately below this plate, the ...

7 Which type of underfloor heating (mono plate or compact plate) would you use for the following rooms? Say why.

1 the living room in a first floor flat
2 the living room in a house without a cellar
3 a room above a loggia (recessed balcony)
4 a room above a cellar

8 Writing

Write a detailed description of a compact plate floor. Include all the parts in the drawing.

9 Talking

Use the list of advantages and other information to give a product presentation of the Monopex underfloor heating system. Add any ideas of your own.

Practising grammar

Remember: Passive → Begleitgrammatik, Abschnitt 15, Seite 157

1 English **is spoken** all over the world.
2 Some floors **are** still **being put in**.
3 This heating system **was made** in Germany.
4 Building work **has been delayed** by bad weather.
5 The heating tube **will be delivered** next week.
6 The heating tube **is not made** of metal, but plastic.
7 **Will** the equipment **be installed** by the supplier?

1 **Put the verbs into the correct passive tense.**

1 They have promised that the flooring … (deliver) by the end of next week.
2 A lot of money … (invest) in new machines since January.
3 … (the new lathe/install) by our own technicians?
4 The agency workers … (pay) in cash every Friday.
5 Our offices … (completely + rebuild) after they … (damage) by fire two years ago.
6 During renovation, all the wooden floors … (remove) and … (replace) by concrete ones.
7 At our firm, a meeting … (hold) to review the week's work every Friday afternoon.
8 … (all products/test) at final inspection or … (only spot checks/carry out)?

2 **Tom, a customer service technician, wrote these notes about problems with a heating system. Write them as a report using the passive.**

1 15. March: Carr & Co. asked us to give presentation of new heating system
2 26. March: placed order for system
3 10. April: delivered system on time.
4 11. April: Carr & Co. carried out inspection of consignment
5 15.–25. April: Carr & Co.'s own contractor installed system
6 26. April: tested system; didn't reach specified temperature

On March 15th, we were asked by Carr & Co. to give a presentation of our new heating system. The company was evidently satisfied because an order for …

Remember: Passive with modals → Begleitgrammatik, Abschnitt 16, Seite 157

1 The user's manual **must be read** by all machine operators.
2 This system **can be installed** over the whole floor area.
3 The heating **should be delivered** next week.

3 **Complete the signs or notices with passive forms.**

Examples:
- The fax machine must/switch on at all times.
  The fax machine *must be switched on* at all times.
- Cars should/not park near these doors.
  Cars *should not be parked* near these doors.

1 This door must/keep free at all times.

2 All fire equipment should/check weekly.

3 Personnel files may/not remove from this office.

4 A hard hat must/wear on the construction site.

5 This seal should/not break except by a technician.

6 Refreshments can/obtain from the canteen after 9 o'clock.

7 The red button may/only/press in an emergency.

8 Security staff should/call to all accidents.

Warm-up

Tr 24 **First match the photos A–H with a word or expression from the box. Then listen to the sounds 1–12 and link them to the photos. Be careful. Four of the sounds do not fit.**

bicycle	horse	jet aircraft	scooter
bus	in-line skates	rowing boat	tram

A

B

C

D

E

F

**Which of the means of transport in the photos do you often/ sometimes/ never use?
What about other members of your group?
Do a class survey.**

G

H

A DESCRIBING A PROCESS: FUEL CELLS

Electric taxi in car-free Zermatt

Hybrid London bus

BMW New Hydrogen 7 luxury sedan

Before reading the text on page 108, make sure you understand these words and expressions:

conversion device	*Umwandlungseinrichtung*
run down	*sich entladen*
continuously	*kontinuierlich*
compete	*konkurrieren*
electricity-generation plant	*Elektrizitätswerk*
advantage	*Vorteil*
direct current voltage	*Gleichspannung*
appliance	*Gerät*
application	*Anwendung*
disadvantage	*Nachteil*
be solved	*gelöst werden*
precious metal catalyst	*Edelmetallkatalysator*
durable	*beständig*
deteriorate	*sich verschlechtern*
volatile	*flüchtig*
fuel delivery system	*Kraftstoffförderung*
impact-resistant	*stoßfest*

A fuel cell is an electrochemical energy conversion device that converts hydrogen and oxygen into electricity and heat. Although it is technically similar to a battery, a fuel cell can never run down because the chemicals it needs are continuously fed into it.

When fully developed, fuel cells will compete with many other types of energy conversion technologies, including electricity-generation plants, internal combustion engines and the batteries in mobile phones and notebooks. Their huge advantage is, though, that they produce no harmful emissions, just heat and water.

In addition, a fuel cell produces a direct current voltage that is suitable not only for electric motors, but also for lighting and many electrical household appliances from cookers to electric toothbrushes.

When you look at all these applications, it is no surprise that there are several different types of fuel cells. These are normally classified by the type of electrolyte they use. Some types are suitable for use in generation plants; others are more useful for powering small portable devices and cars.

However, fuel cells have several disadvantages that still have to be solved. To begin with, energy generated by fuel cells is very expensive – too expensive for general use, in fact. At present, one kilowatt (kW) of energy produced by an internal combustion engine (ICE) for $35 costs $110 when produced by a PEM fuel cell using a precious metal catalyst like platinum.

Secondly, fuel cells are not very durable, much less so than an ICE, say. The reason is that fuel cells deteriorate relatively quickly when frequently turned on and off, as in a road vehicle or household appliance.

Finally, there is the question of safety. Pure hydrogen is extremely volatile. There is a danger of explosions, particularly in traffic accidents. This means that the hydrogen tanks and fuel delivery systems in road vehicles have to be impact-resistant, adding to both weight and cost.

1 **Answer the questions.**

1. Say what the purpose of a fuel cell is.
2. Point out the big advantage of a fuel cell over a battery.
3. List the fuel cell applications that are mentioned in the text.
4. Say how fuel cells are generally classified.
5. Point out why fuel cells are attracting so much interest at present.
6. Describe the problems that must be solved to make fuel cells suitable for general use.

2 Tr 25

Work with a partner or in small groups. Study the drawing and try to complete the process description with the missing words. Then listen to the CD to check your answers and fill in any empty gaps.

How proton exchange membrane fuel cells (PEMs) work

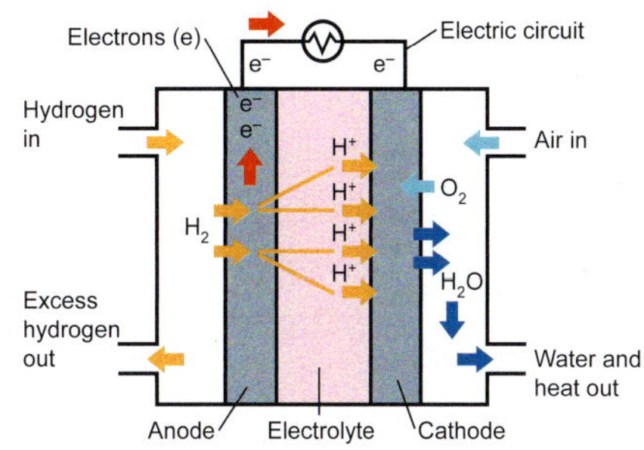

In a PEM fuel cell, … [1] (H_2) atoms enter the cell on the … [2] (negative) side. A chemical reaction with a catalyst, usually … [3] (Pt), on the porous … [4] separates the electrons (e^-) from the hydrogen atoms. These negative … [5] become the direct … [6] (DC) that powers the … [7], a light bulb, say. The hydrogen ions (h^+) now have a … [8] electrical charge.

Meanwhile, … [9] from the air (O_2) enters the cell on the … [10] (positive) side. In a PEM fuel cell, oxygen … [11] combine with the negative electrons (e^-) returning from the load and the positive … [12] (h^+) that have passed through the electrolyte. The result of this … [13] is pure … [14] (H_2O) and some excess … [15].

3 **Single fuel cells designed for use in vehicles generate less than 1.16 volts, far too little to power a car. Look at the illustration. How can the output of fuel cells be increased?**

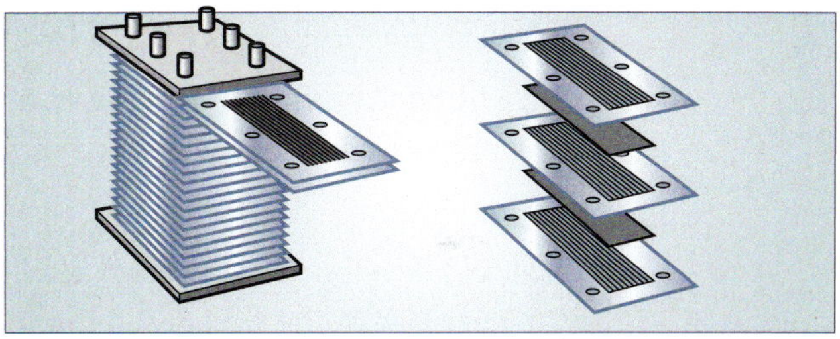

4 Tr 26

Listen to the radio talk about problems with fuel cells. Use a grid like the one below to note down the speaker's main points. Make sure you understand the words and expressions in the list first.

Word list

limitless	*grenzenlos*
be obtained	*(hier:) gewonnen werden*
contaminated	*verunreinigt*
be purified	*gereinigt werden*
on site	*vor Ort*
be distributed	*verteilt werden*
impracticable	*unausführbar*
property	*(hier:) Eigenschaft*
be stored	*gelagert werden*
pressure	*Druck*

Production problems	
Distribution	
Problems in vehicles	
Pollution	

B Electric vehicles (EVs)

Most people agree that electric motors are far superior to internal combustion engines (ICEs). Firstly, they are far more efficient. Whereas an ICE turns only about 35% of the energy it produces into kinetic energy, i.e. movement, an electric motor uses over 90% of its energy to move the vehicle. Secondly, electric motors produce no harmful emissions, while ICEs are an ecological disaster. So much is agreed. The discussion now is about how to power the electric motor, with a hydrogen fuel cell or with a battery?

The argument against batteries is certainly not their cost as they are many times cheaper than a hydrogen fuel cell. In modern batteries, efficiency is not a problem, either. A battery delivers some 90% of the electricity it stores to the motor, making it over twice as efficient as a fuel cell.

Why, then, aren't batteries used to power electric cars?

The traditional answer to this question is that batteries need recharging too frequently, giving battery-powered cars only a very short range.

This was certainly once the case. Look at General Motors' legendary EV-1, for example, which was developed in the early 1990s to meet California's strict "zero-emissions law" for at least 10% of the vehicles in the state.

The EV-1 was way ahead of its time. It had a top speed of very nearly 290 km/h and an acceleration from zero to 100 km/h of only 7.8 seconds. However, the problem was that the EV-1 only had a range of 160 kilometres between charges, much too short in a country as big as the USA. For this reason, the car did not sell well and GM was forced to stop production.

However, advances in battery technology have already changed this situation completely, and further advances are likely. The use of nickel-metal hydride batteries instead of lead-acid batteries more than doubles the range between recharging, in the case of the EV-1, for example, to around 340 kilometres. Furthermore, fitting advanced lithium-ion batteries would increase this range yet further to 480 kilometres and more, according to how the car is driven. This is easily long enough for most purposes.

This is why some experts think that fuel cells could become out of date before their numerous technical problems have been solved.

5 **Choose the correct alternative, a) or b).**

1 Most people think that
a) hybrids are superior to 100% electric cars.
b) electric motors are better than ICEs.

2 An ICE
a) wastes about 65% of the energy input.
b) is more efficient than an electric motor.

3 The big controversy now is
a) which power source should drive electric motors.
b) the advantages of hybrids over electric cars.

4 Batteries are over twice as
a) efficient as fuel cells.
b) heavy as fuel cells.

5 Lead-acid batteries
a) must often be renewed.
b) do not have a long range between recharges.

6 The EV-1 was developed to reduce
a) America's dependency on imported oil.
b) car emissions.

7 The best type of battery for cars is the
a) nickel-metal hydride battery.
b) lithium-ion battery.

8 Some experts think
a) that further advances in battery technology are unlikely.
b) the technical problems of fuel cells will never be overcome.

6 **a) Rewrite the sentences in the passive. Only use a *by*-agent if really necessary.**

General Motors introduced the EV-1 in the early 1990s.
The EV-1 was introduced by General Motors in the early 1990s.

1. They did this to meet California's zero-emissions law.
2. The state government designed this law to reduce air pollution.
3. Since then, they have brought in several other laws to protect the environment.
4. For example, you will have to fit a microfilter to all diesel engines from 1 January 2009.
5. While Europeans often criticise the USA about climate change, in fact individual states are passing very strict laws to protect the environment.
6. Some states have abolished sales tax (*Mehrwertsteuer*) on bikes.
7. Another idea is that shops can sell only low-energy light bulbs and rechargeable batteries.

b) Put the verbs into the correct passive form of the simple present or simple past.

The Eurotunnel – or Channel tunnel as it ... [1] (sometimes + call) – ... [2] (open) in 1994. The tunnel ... [3] (make) up of two parallel rail tunnels, one in each direction, that ... [4] (separate) by a smaller service tunnel. The service tunnel, which ... [5] (connect) to the two rail tunnels every 375 metres, is a road tunnel that ... [6] (only + use) by specially-designed vehicles.

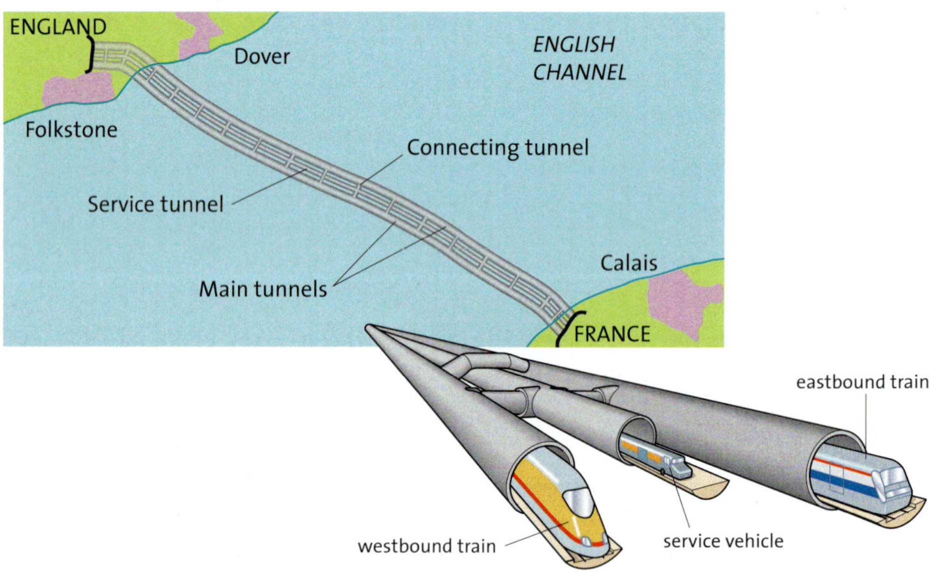

These multi-purpose vehicles ... [7] (design) and ... [8] (build) in Germany. They ... [9] (power) by electric motors that ... [10] (driven) by rechargeable batteries. Their bodies ... [11] (can change) into ambulances, fire engines or buses for use in emergencies.

This happened in November 1996 when one of the tunnels ... [12] (have to close) when a truck on a train caught fire. Some special vehicles ... [13] (change) into fire engines and others into buses. The fire ... [14] (quickly + put out) and a lot of passengers ... [15] (carry) to safety through the service tunnel. Everybody ... [16] (save) and nobody ... [17] (injure).

7 Tr 27

Study the drawing and complete the text with words and expressions from the box. Look at the word list first. Then listen to the CD to check your answers and fill in any empty gaps.

Word list

consist of	*bestehen aus*
current	*Strom*
enable	*ermöglichen*
voltage	*Spannung*
be released	*losgelassen werden*
pulse	*pulsieren (lassen)*
maintain	*(hier:) erhalten*
in succession	*nacheinander*
travel	*(hier:) Weg*
occur	*stattfinden*

batteries	on/off switch
components	pedal
controller	potentiometer
current	pulses
electric vehicle	rate
electricity	signal
given speed	travel
in succession	value
motor	voltage

How a DC automotive power unit works

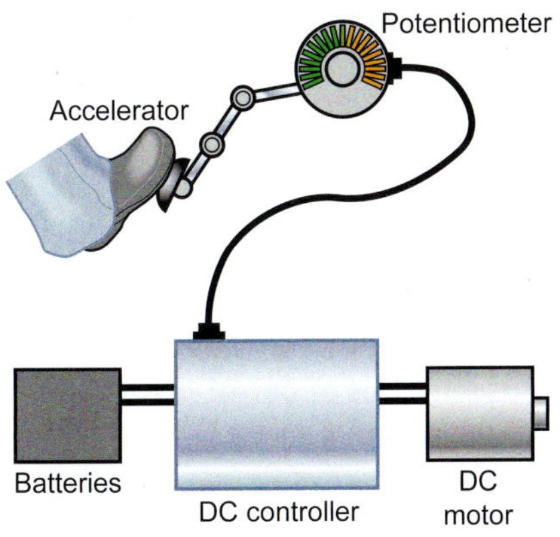

The power unit of a DC ... [1] consists of three main ... [2]: a DC motor, a DC ... [3] and a "battery array" of eight 12V ... [4] in series, giving 96V in all.

The controller takes ... [5] from the battery array and sends it to the ... [6]. The accelerator pedal in the car is linked to a ... [7] (variable resistor) that sends a ... [8] to the controller that tells it how much current to deliver. The potentiometer enables the controller to deliver any ... [9] to the motor from zero to 96V, and any ... [10] in between.

The most basic form of DC controller would be a simple ... [11] connected to the accelerator pedal. When the pedal is pushed down, ... [12] flows and when the pedal is released, it stops. However, this means that the ... [13] would have to be pushed down and released many times to pulse the motor in order to maintain a ... [14].

Although technically possible, the vehicle would be extremely difficult to drive. For this reason, the controller ... [15] the motor as well.

How does this work?

If the pedal is pushed halfway down, the controller switches the motor on for half the time and off for half the time ... [16]. Similarly, if the pedal is pushed down 25% of its ... [17], then the controller pulses the current so that the motor is on 25% of the time and off 75% of the time. As pulsing occurs in most electric vehicles at a ... [18] of 15,000 times per second, the on/off action of the motor is not felt by people in the car.

8 **Work in small groups. Study the drawing and explanation. Then describe how an AC electric power unit works. Use the text about DC power units as a model.**

How an AC automotive power unit works

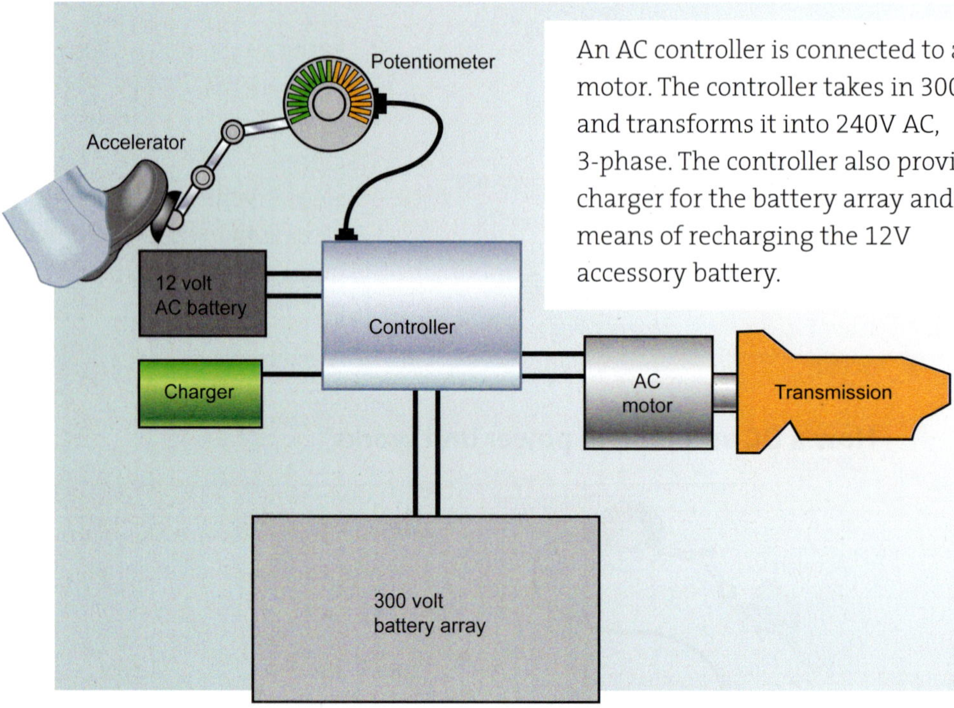

An AC controller is connected to an AC motor. The controller takes in 300V DC and transforms it into 240V AC, 3-phase. The controller also provides a charger for the battery array and a means of recharging the 12V accessory battery.

9 **Work with a partner. Think about the pros and cons of petrol-driven cars on the one hand and electric cars on the other. Copy the table on page 115 and complete it with material from the box. Then compare your result with those of other members of your class.**

comfortable
easy and quick to refuel
good weight-carriers: up to 500 kg
harmful emissions: CO, CO_2
high fuel consumption: approx 11.5 l/100 km (average)
highly efficient
long range: up to 700 km
narrow speed range: approx 2 to 100 km/h (max.)
no harmful emissions
poor weight-carriers: approx 160 kg
short range: approx 150 km
slow recharging: approx 8 hours
uncomfortable for big adults
wide speed range: 4 to 200 km/h

Petrol-driven cars		Electric cars	
Pros	**Cons**	**Pros**	**Cons**
long range: up to 700 km	...	...	short range: ca. 150 km
...	...	...	...
...	...	...	...

10 Tr 28

First read through the ten statements below. Then listen to the extract from Ed Jackson's radio phone-in show about cars and climate change. After that link the statements to the caller. They are not necessarily in the same order as in the text, of course. Be careful. Two of the statements do not fit at all.

1 Does not think car owners need change their lifestyle at all.
2 Doubts getting rid of a car helps, as somebody else will probably use it.
3 Doubts that wind farms and solar energy will generate enough power.
4 Feels we can keep our cars, but they should be as eco-friendly as possible.
5 Has got rid of one of two cars and kept the smaller one.
6 Is sure that a technical solution to climate change will be found.
7 Points out that radical anti-climate change policy could destroy jobs.
8 Says climate change is not taking place at all, just natural changes in the weather.
9 Thinks we are destroying our towns and countryside with roads and car parks.
10 Thinks we must change our lifestyle because of climate change.

11 Talking

Which of the callers do you agree with most? What would you be willing to do to reduce carbon emissions – or aren't you really that worried? Give reasons.

12 Writing

Materials used in cars: percentage of total weight

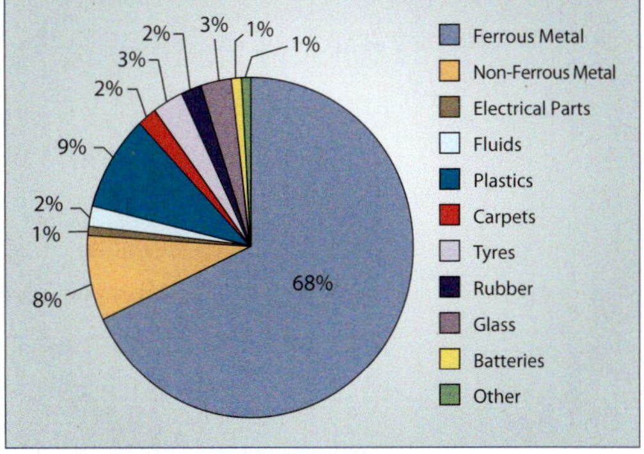

In the UK, there are over 30 million cars and light vans, which makes one for every two people in the population as a whole.
Each year, just over two million new vehicles replace just under two million old ones, giving a net increase of about 50,000 vehicles per year. But what happens to the old cars? The problem of people leaving old cars on the sides of the road and in the countryside has become so bad that the EU is now trying to make car manufacturers provide a free take-back service.

Use the information in the pie chart to write a report about car disposal.

Practising grammar

Remember: Adjectives and adverbs → Begleitgrammatik, Abschnitt 17, Seite 158

1 A **big** advantage of fuel cells is that they produce no **harmful** emissions.
2 A **powerful** argument for **electric** motors is that they are **clean** and **efficient**.
3 Fuel cells deteriorate **quickly** when they are switched on and off **frequently**.
4 The test did not go **well**. The machine ran far too **fast** and then cut out **early**.

1 **Add the adjectives in brackets to the sentences. Use the adverb form where necessary. Be careful. Some are irregular.**

1 Henry's a … 1 technician, and he's repaired the machine … 2 … 3. (*good, real, good*)
2 You're talking too … 1 and … 2. Could you speak more … 3 and … 4, please? (*fast, loud, slow, quiet*)
3 Stop work … 1! The instructions say that it's … 2 to check all components … 3 before using the machine for the … 4 time. (*immediate, important, careful, first*)
4 Peter's not very … 1 with his … 2 car. Its fuel consumption is far too … 3 and so it's very … 4 to run. It … 5 used … 6 eight litres per hundred kilometres, which is … 7 in such a … 8 engine. (*happy, new, high, expensive, recent, near, absurd, small*)
5 The sales of hybrids have risen … 1 in the UK this year although car companies increased their prices … 2. … 3, the credit crunch (*Kreditknappheit*) hasn't had a … 4 effect on sales, either. (*slight, sharp, surprising, big*)
6 … 1, Jack … 2 saw that the motor was running … 3 and … 4 switched it off. That was … 5 because we could have had a … 6 fire. (*lucky, quick, rough, immediate, lucky, bad*)

2 **Read the car ad and choose the correct alternative, a) or b).**

The Mazda MX-5

Motoring is a lot more more than getting from A to B as **a)** quick / **b)** quickly 1 and **a)** safe / **b)** safely 2 as possible; it's there to give you a **a)** real / **b)** really 3 thrill as well. And in your **a)** new / **b)** newly 4 developed Mazda MX-5, a **a)** terrific / **b)** terrifically 5 motoring experience is **a)** absolute / **b)** absolutely 6 guaranteed. The Mazda MX-5 is **a)** official / **b)** officially 7 the world's Number One open-top sports car. And this isn't **a)** surprising / **b)** surprisingly 8 when you look at the **a)** complete / **b)** completely 9 redesigned MX-5's combination of **a)** high / **b)** highly 10 advanced engineering with that **a)** authentic / **b)** authentically 11 "wind-in-the-hair" sports car feel. Where we've been able to reduce weight without loss of safety, we've done so. The result is a **a)** light / **b)** lightly 12 but **a)** robust / **b)** robustly 13 sports car, with **a)** perfect / **b)** perfectly 14 handling and performance to be **a)** true / **b)** truly 15 proud of.

Remember: Comparisons → Begleitgrammatik, Abschnitt 18, Seite 159

1. Electric motors are **cleaner** and **more efficient** than diesel engines, say.
2. Sure. In fact, of all power sources, electric motors are the **cleanest** and **most efficient**.
3. The test went **better** today. It ran **more slowly** and cut out **later** than it did yesterday.

3 **Use the information in the table to write sentences comparing battery-powered motors with ICEs.**

++ excellent / + good / O average / – bad / –– terrible

	Efficiency	Cost	Reliability	Range	Load capacity	Safety
Motors	++	O (batteries)	+	–	––	+
ICEs	––	+	O	+	++	+

The efficiency of motors is excellent in comparison with that of ICEs, which is terrible. Looking at cost, motors are more expensive than ICEs because of the batteries, but on the other hand they are more reliable. ...

4 **Translate the sentences into English.**

1. Natürlich sind die meisten technischen Übersetzungen schwieriger als „normale" Texte.
2. Ich finde, dass der neue Golf viel besser aussieht als das alte Modell.
3. Leider werden Benzin und Diesel so teuer, dass wir unser Auto nur für absolut notwendige Fahrten benutzen können.
4. Wir versuchen, unsere Gebrauchsanweisungen (*user manuals*) so klar und einfach wie möglich zu schreiben, aber manchmal ist das nicht einfach.
5. Heute ist der Betriebsleiter (*plant manager*) nicht hier. Zum Glück ist er aber nicht ernsthaft krank.
6. Es mag sein (*it could be*), dass Brennstoffzellen keine umweltschädlichen Emissionen produzieren, aber sie benötigen viel Elektrizität von „dreckigen" Kraftwerken in der Herstellung.

Warm-up

After years of argument about whether the climate is changing at all – and, if it is, why – expert opinion now says that the effect of climate change on the global economy is by far the most urgent threat to life on Earth as we know it. In other words, the global warming caused by climate change is no longer a question of "if" and "why", but of "when" and "how". Read on to find out more.

1

2

3

4

5

6

Do the assignments.

1 Use the numbers 1–6 to link the photos to the captions A–F.

A Floods and rising water levels are drowning low-lying communities.
B Glaciers and ice-caps are melting much faster than expected.
C First drought and then fire are destroying forests.
D Drought is devastating huge areas of farmland.
E Hurricanes and severe storms are becoming more frequent and more destructive.
F The heat is killing many species while turning others into dangerous pests.

2 Which of the six photos do you find most worrying? Choose one and say why.

What's your take on climate change?

Although opinion surveys show that most people agree that climate change is a reality, there's still a lot of argument about both its causes and its effects.
While a majority think that it is caused mainly by human activity, particularly carbon emissions, a fairly large minority say that it has largely natural causes. They point out that there have always been big fluctuations in the world's climate.
And people also disagree strongly about its effects. Surveys suggest that many people think that a lot of what we hear is just "media hype" (*Medienrummel*). They feel that, in time, we will find technical answers to all the problems of climate change. Some observers even think that there will be some winners, at least in colder regions such as Russia, Canada and Scandinavia.

3 First, read the opinion survey on climate change below and make sure you understand the eight statements. Then do the survey by choosing the alternative a-c that is nearest to your own opinion.

Online Questionnaire | Climate Change Explorer
http://www.climatechangeexplorer.org.uk/onlinequestionnaire
Climate Change Explorer — Please Login or Register

An online opinion survey on climate change
We are interested to hear your opinion on climate change. Do our online survey by clicking the opinion that you agree with most.

1 The world's climate is definitely changing.
a I agree. **b** I disagree. **c** I don't know.

2 Climate change is caused mainly by humans.
a I agree. **b** I disagree. **c** I don't know.

3 Climate change is a natural phenomenon.
a I agree. **b** I disagree. **c** I don't know.

4 I have become much more aware of climate change recently.
a I agree. **b** I disagree. **c** I don't know.

5 Everybody can do something to fight climate change.
a I agree. **b** I disagree. **c** I don't know.

6 The poor will suffer. The rich will buy their way out of trouble.
a I agree. **b** I disagree. **c** I don't know.

7 Climate change will have some winners, Siberia, for example.
a I agree. **b** I disagree. **c** I don't know.

8 The climate is changing all right, but its effects are very exaggerated.
a I agree. **b** I disagree. **c** I don't know.

Submit

4 Copy the table below into an exercise book and use it to compile the questionnaire results for your class.

	a	b	c
1	*19*	*8*	*3*
2			

5 Complete the statement below to report the opinion of the whole group.

Generally, ... respondents agree that climate change is ... However, a minority of ... disagree with this and ... respondents don't know. As far as the causes of climate change go, ... think that it is caused by..., while ... respondents ... natural causes. Here, too, ... people gave no opinion either way. Turning to the effects of ..., ... people feel that these ..., and ... even think there ... as well as losers. ... people think that we can all ...

A THE DEBATE ABOUT GLOBAL WARMING IS OVER – BUT WHAT NOW?

Scientific opinion is clear: global warming caused by humans is real, and because of it the world's climate is changing. We will certainly be able to correct some of the **causes** of climate change by technical means – by using renewable energy sources and improving the insulation of our homes, for example. We will certainly be able to fight some of the **effects** of climate change by technical means, too – by building higher and better flood defences, for example. But let's not kid ourselves. Technology can only help to protect us from the worst consequences of our own behaviour if we change our lifestyle as well.

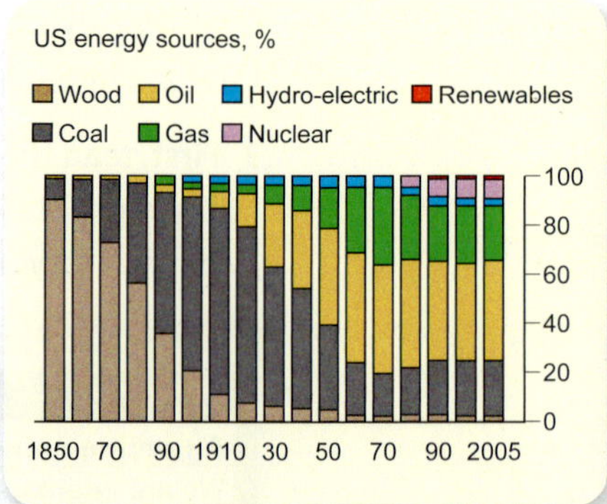

A tour of the greenhouse

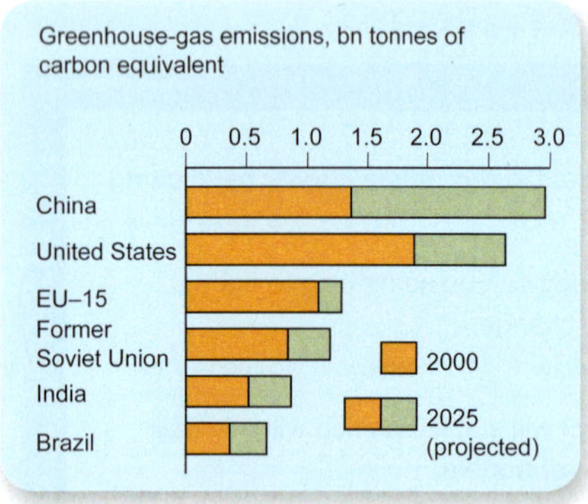

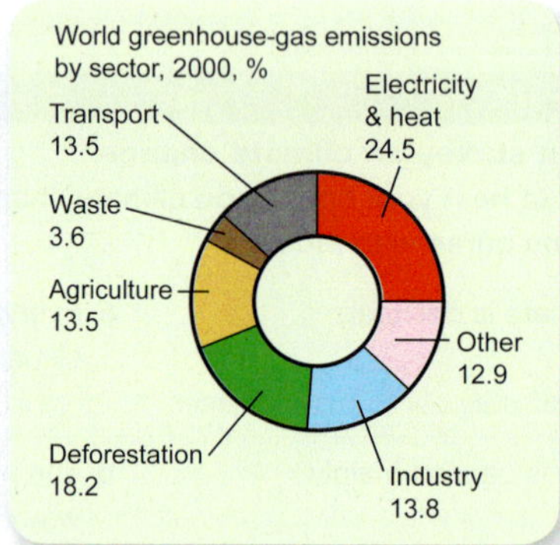

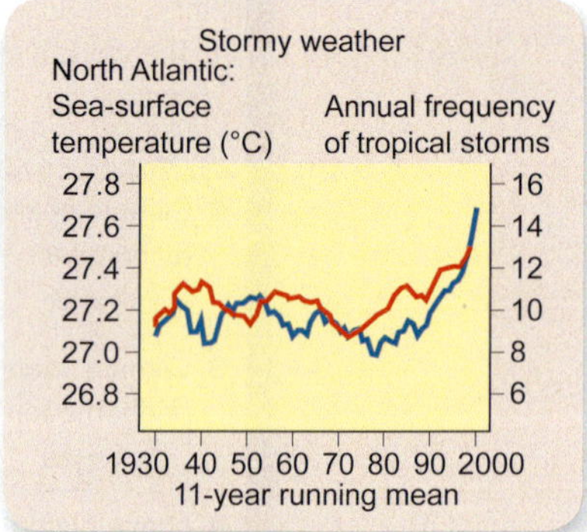

1 Answer the questions.

1. Compare global greenhouse-gas emissions in 2000 with those of 2025. What do you notice?
2. Which sector is responsible for emitting most carbon? Why is this the case?
3. Can you say – or guess – why agriculture is responsible for emitting as much carbon as transport and nearly as much as industry?
4. How many energy sources are used in the USA? What huge change has taken place since 1850?

2 Replace the German words or expressions with their English equivalents from the box.

annual frequency	deforestation	of fossil fuels
average	emissions	pie chart
bar chart (2×)	energy sources	projected
billion	environmental pollution	surface temperature
burning	generation	
by various countries	line graph	

1 The *Liniendiagramm* [1] relates the *Oberflächentemperatur* [2] of the sea to the *Jahreshäufigkeit* [3] of tropical storms. The horizontal axis shows the *Durchschnitt* [4] from 1930 to 2000.
2 The *Balkendiagramm* [1] shows how many *Milliarden* [2] tonnes of carbon *Emissionen* [3] were produced *von verschiedenen Ländern* [4] in 2000, and will be produced by 2025. This last figure is *hochgerechnet* [5].
3 The *Kreisdiagramm* [1] shows world carbon emissions by sector. It presents seven main sources of *Umweltverschmutzung* [2] through carbon, the most important being electricity and heat *Erzeugung* [3] and *Abholzung* [4].
4 The *Säulendiagramm* [1] shows how US *Energiequellen* [2] developed between 1850 and 2005. Basically, the USA changed from the *Verbrennung* [3] of wood to the use *von fossilen Brennstoffen* [4] like oil and gas.

3 Writing

Work in small groups. Use the information given in the charts on page 120 to write a report on global warming. Add any information or ideas of your own.

Describing graphs and charts

■ Line graphs

The line graph deals with/shows the relationship between ... and ... from ... to ...
The graph shows/illustrates a steady/sharp increase/rise/decrease/fall in ... over the last two/... years/...
The line climbs/rises gradually/slowly/sharply/steeply/fast and peaks at/reaches a peak at .../a maximum value of ...
The line falls gradually/slowly/sharply/steeply/fast and bottoms out at .../at a minimum value of ...

■ Bar charts and pie charts

The bar chart compares ... from ... to ...
The pie chart deals with/shows the ...
The pie chart is divided into five segments showing ... in .../during the period from ... to ...
The pie chart shows/makes clear the ... in ...

- A line graph shows changes in value (***Wert***) over time.
- A bar chart is generally used to compare values at a point in time or over a period of time.

We use "bar chart" for both **Säulen**- and **Balkendiagramme**.

- The segments of a pie chart show a given value in relation to the whole.

4 Talking

Work in small groups. Discuss ways of reducing the carbon that is escaping into the atmosphere and causing global warming. Then report back to the class as a whole.

■ Here's some help:

– Brainstorm ways of reducing carbon, for example ...

- ☐ *banning standby switches*
- ☐ *using low-energy light bulbs*
- ☐ *replacing conventional power plants with nuclear power plants*
- ☐ *making greater use of solar-, water-, wind-power*
- ☐ *taxing aviation fuel*
- ☐ *improving/subsidising public transport*
- ☐ *increasing petrol/diesel tax*
- ☐ *improving inland waterways/barge traffic*
- ☐ *introducing car-free Sundays*
- ☐

Making suggestions

– I suggest that we ban the use of standby switches/...
– I think we should build cycle tracks/...
– In my opinion/view we should ...
– Isn't it possible to tax aviation fuel/...

– How/What about using barges instead of trucks/...?	***about*** + -***ing*** form
– What's wrong with building more nuclear power plants/...?	***with*** + -***ing*** form
– If we build safe cycle tracks, people will use bikes more.	***if***-sentence, type I
– If we used nuclear power/..., we would produce less carbon.	***if***-sentence, type II

Rejecting suggestions

– That's no good because of the cost/...
– I disagree. I think that idea is impracticable/...
– Yes, but people love their cars/... I'm sure they would reject that idea.
– Okay, but don't forget that .../remember that ...

A telephone opinion poll

5 Listen to the radio report.

Tr 29

Before beginning ...

a) read the questions below so that you know what information to listen for and

b) make sure you know the meaning of these words and expressions: *poll, respondent, promising, public awareness, gasoline* (AE).

Copy the questions and tables into your exercise book. Then complete them with the missing information.

■ "Has the world's ...¹ been rising over the past 100 years?"

Year	Yes	No	Don't know
2007	... ² %	... ³ %	none
1997	... ⁴ %	30 %	... ⁵ %

■ "Is ... ⁶ already a serious problem or is it a problem for the future?"

Year	Already a problem	A problem for the future	Don't know
2007	... ⁷ %	54 %	... ⁸ %

■ "Do you think the federal government should ...

... increase ... ⁹ on gasoline so people consume less of it?"	**Yes**: 28 %
... increase taxes on ... ¹⁰ so people consume less of it?"	**Yes**: ... ¹¹ %
... subsidise the development of ... ¹² sources?"	**Yes**: 87 %

How would you and other members of your group answer the questions in the poll? Do a class survey and compare the results with the results that Poll America got.

B GEOTHERMAL ENERGY

One of the most promising ways of reducing the CO_2 emissions that are a main cause of global warming is to use less carbon-intensive fossil fuels like oil and coal, and more renewable energy such as the wind and the sun. (Nuclear power can also reduce our use of fossil fuels, but is it wise to start an engine that you can't switch off? Most people say "No".)

However, some parts of the world are luckier than others. They have a renewable energy source that is both close to home and relatively easy to reach – the heat at the earth's core.

6 Read the text about the earth's interior. Say what a–d are.

Geothermal energy is generated in the earth's core, about 6,500 kilometres below the surface. Temperatures hotter than the sun's surface are continuously produced by the slow decay of radioactive particles, a process that happens in all rocks.

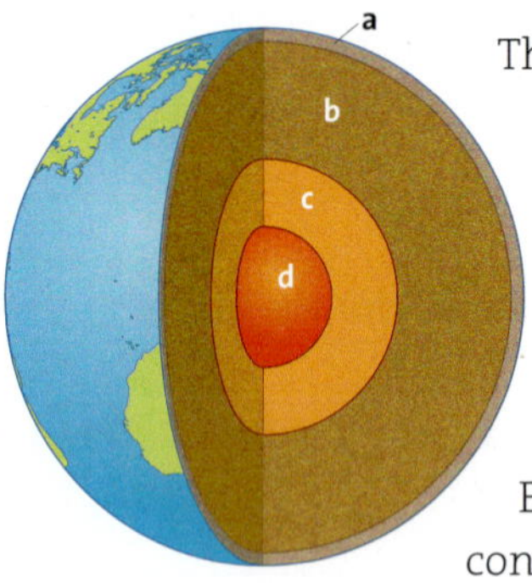

The earth has four layers, two of which make up its core. At the very centre of the core there is a solid mass called the iron core. Around the iron core is the outer core made of very hot molten rock, called magma. The whole core has a diameter of about 3,500 kilometres. The mantle that surrounds the core is just under 2,900 kilometres thick. It is made up of a mix of molten magma and solid rock. The Earth's outer layer is called the crust, the land that forms the continents and ocean floors. It is about five to eight kilometres thick under the oceans and about 25 to 55 kilometres thick on the continents.

The earth's crust is broken into pieces called "continental plates" or simply "plates". Magma comes close to the earth's surface near the edges of these plates. This is where earthquakes can occur and volcanoes are found. The lava that erupts from volcanoes is partly magma. Deep underground, the rocks and water absorb the heat from this magma. The temperature of the rocks and water becomes progressively higher as you go deeper underground.

People around the world use geothermal energy to heat their homes and to produce electricity by digging deep wells and pumping the heated underground water or steam to the surface. Most geothermal energy is deep underground with no visible clues showing above ground. However, geothermal energy can sometimes find its way to the surface naturally in the form of volcanoes and fumaroles (holes where volcanic gases are released), hot springs and geysers such as those in Iceland.

As the most accessible geothermal resources are found along major plate boundaries, most occur in an area called the Ring of Fire along the edge of the Pacific Ocean.

When magma comes close to the surface, it heats ground water found trapped in porous rock or water running along fractured rock surfaces and faults. Large areas of such heated water are called geothermal reservoirs. Geologists use different methods to look for geothermal reservoirs. Drilling a well and testing the temperature deep underground is the only way to be sure a geothermal reservoir really exists.

7 Link a sentence part from list A, 1–10, to one from list B, a–j, to make sensible statements.

A

1. Geothermal energy is generated
2. The slow decay of radioactive particles
3. The earth consists of four layers of matter,
4. The earth's mantle consists of
5. The earth's crust is very much thicker on land
6. The earth's crust is made up
7. Earthquakes take place when magma
8. Rocks and water in the crust absorb
9. Geothermal energy appears naturally
10. The most easily accessible geothermal resources

B

a about 6,500 km below the earth's surface.
b a mix of molten magma and solid rock.
c are found in the Ring of Fire around the Pacific.
d as volcanoes and geysers, for example.
e heat from the mantle deep underground.
f nears the surface at the edges of continental plates.
g of huge pieces called "continental plates".
h takes place in all rocks.
i than it is below the oceans.
j two of which make up its core.

1 *e*
2 ...

8 Find a technical term in the text to replace the terms of general meaning in the list.

1. heat from the middle of the earth
2. the earth's outer skin
3. the middle of the earth
4. liquid rock in the middle of the earth
5. the very centre of the earth
6. the bottom of the sea
7. pieces of the earth's surface
8. liquid rock that comes out of volcanoes
9. place where hot steam or water comes out of the earth naturally
10. rock with tiny (*winzig*) holes in it

9 **Explain these words and expressions in German. Then use German to write a paragraph about geothermal energy.**

1 geothermal energy	**3** magma	**5** lava	**7** ring of fire
2 layer	**4** continental plate	**6** geyser	**8** geothermal reservoir

Geothermal power plants

There are three types of geothermal power plants that convert geothermal energy into electricity: dry steam, flash and binary cycle. The type of plant is determined by the energy source, steam or water, and its temperature.

10 **Link the three drawings below to the correct part of the text.**

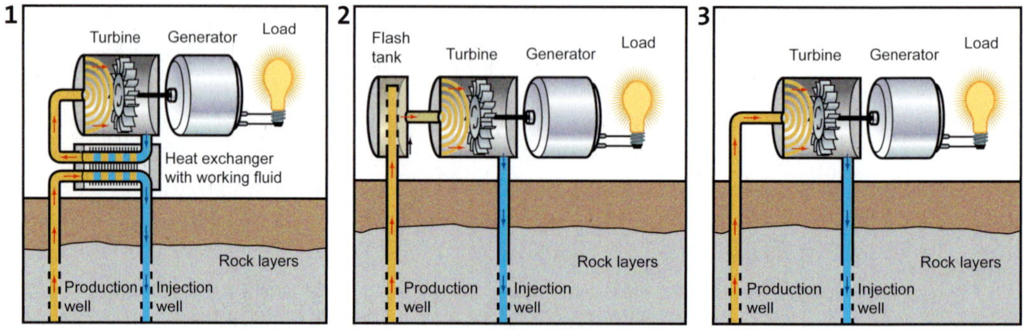

A

Dry steam power plants use natural steam from geothermal reservoirs as it emerges from the earth. The steam is transmitted directly to turbines driving generators to produce electricity. Dry steam power plants are both the technically simplest and the cheapest type of geothermal power plant, but unfortunately there are very few parts of the world where natural steam can be harnessed in this way.

B

For this reason, flash steam plants are the most common type of geothermal electricity generation system. They use water at temperatures greater than 182°C that is pumped under high pressure to the generation plant at the surface. Hot water is sprayed into a tank held at a much lower pressure than the water, causing it to instantly vaporise, or "flash." The vapour then drives a turbine, which powers a generator. If any water remains in the tank, it can be flashed again in a second tank to extract even more energy.

C

Binary cycle geothermal power generation plants differ from both dry steam and flash steam systems in that the water or steam from the geothermal reservoir never comes into contact with the turbine unit.

11 **Look at the relevant drawing on page 126 and explain how a binary cycle power plant works.**

In binary cycle power plants, moderate-temperature (< approx 200 °C) geothermal water and a secondary (hence, "binary") liquid with a much lower boiling point ...

12 **Giving a presentation**

Work in groups of 3 or 4. Use the outline to prepare a presentation on the energy crisis, using information from this unit and any ideas of your own. (The internet sites could help you here.) When you are satisfied with your presentation, appoint a member of the group to present your findings and views to the rest of the class. Wherever possible, make use of illustrations, etc. to support what you say and make your presentation more "audience-friendly".

Outline: Solving the energy crisis

Introduction
Describe current situation: increasing fuel shortages/increasing demand (China, India, Brazil, ...) → sharply rising prices → search for new/cheaper energy sources → (among other things) "food for fuel" programmes → civil unrest among poor → ...

Development I
Technical answers: renewable energy (name sources); nuclear energy; more efficient energy transformation, e.g. replacement of internal combustion engines; low-energy devices, e.g. lighting; ...

Development II
Lifestyle answers: energy conservation; resource recycling; bikes, buses and trains vs. cars and planes; barges vs. trucks; small is beautiful; ...

Conclusion
Your group's answer/recommendations/conclusions

Useful websites
- Energy Information Administration at www.eia.doe.gov
- Energy Saving Trust, UK at www.energysavingtrust.org.uk
- Friends of the Earth at www.foe.org (USA), www.foeeurope.org (Europe)
- Bund der Energieverbraucher at www.energieverbraucher.de
- O.Ö. Energiesparverband at www.esv.or.at (excellent links)

Practising grammar

Remember: *If*-sentences, types I and II → Begleitgrammatik, Abschnitt 19, Seite 159

1 If you **pump** geothermal water into a flash tank, it **vaporises**.
2 If we **do** nothing about climate change, the situation **will become** even worse.
3 If we **increased** electricity prices, maybe people **wouldn't waste** it.

1 **Complete the *if*-sentences with the missing verb. Look at the examples first.**

Examples: – If I miss my plane, I ... (phone) you.
If I miss my plane, I *will phone* you.

– Nuclear fuel ... (be) the best source of electricity, if it was safe.
Nuclear fuel *would be* the best source of electricity, if it was safe.

1 If we used more renewable energy, we ... (not produce) so much CO_2.
2 We ... (do) much better if we invested more money in training.
3 If the pipelines ... (not be) ready, they can't deliver any gas.
4 If cars are banned from town centres, a lot of shops ... (close).
5 We ... (burn) even more dirty coal if we close down our nuclear power plants.
6 We would use more nuclear power if we ... (have) somewhere to put the waste.
7 If we had more geothermal energy, we ... (build) more geothermal power plants.
8 People won't change to public transport even if petrol price ... (go) up even more.
9 I'll ask Sarah to call you if I ... (see) her in town.
10 If the plant was ready, we ... (open) immediately.

Remember: *If*-sentences, type III → Begleitgrammatik, Abschnitt 20, Seite 160

1 If you **had taken** a train, you **would have arrived** on time.
2 If we **had done** nothing, we **would have been** in an even worse situation.

2 **Form Type III *if*-sentences.**

1 The machine ... (catch) fire if Benny ... (not switch) it off.
2 If the plane ... (take off) on time, we ... (arrive) in time for the meeting.
3 If you ... (deliver) the software as agreed, we ... (win) the order.
4 If Greg ... (not wear) a hard hat, he ... (be) badly hurt.
5 I ... (meet) you at the airport if you ... (tell) me your arrival time.
6 If the pilot ... (follow) instructions, the accident ... (not happen).

3 **Finish translating the sentences into English.**

Examples:

– *Wenn wir die Benzinpreise stark erhöhen, werden die Hersteller sparsamere Motoren entwickeln.*
If we ... sharply, ... more economical ...
If we *increase petrol prices* sharply, manufacturers *will develop* more economical engines.

– *Wenn ich mehr Zeit hätte, würde ich eine Abendschule besuchen.*
If ..., I ... evening classes.
If I *had more time*, I *would go* to evening classes.

1 *Wenn die Ersatzteile heute eintreffen, werden wir sie sofort einbauen.*
If the ... today, we ... immediately.

2 *Wenn wir das Geld hätten, würden wir mehr Arbeiter einstellen.*
If we ..., we ... more workers.

3 *Wenn Sie diesen Hebel betätigen, wird sich die Maschine automatisch abschalten.*
If you ... lever, ... automatically.

4 *Wenn wir die Maschinen richtig warten, werden wir weniger Ausfallzeiten haben.*
If we ... properly, we ... less downtime.

5 *Wenn Züge zuverlässiger wären, würde ich sie sicherlich häufiger benutzen.*
If trains ..., I ... more often.

6 *Wenn Sie gute Deutschkenntnisse hätten, würden Sie die Stelle bekommen.*
If you ... German, you ...

7 *Wenn wir die Computer in dieses Büro stellen, werden wir mehr Steckdosen benötigen.*
If we ..., we ...

8 *Wenn er nicht so schnell gefahren wäre, wäre der Unfall nicht passiert.*
If he ..., the accident ...

4 **a) Form sentences to say what you will do if you ...**

1 ... earn a lot of money.
2 ... don't find a suitable job or lose your job.
3 ... have the chance to spend two weeks anywhere in Europe.

b) Now say what you would do if you ...

4 ... met your favourite pop star or film star.
5 ... lost your way in a strange town.
6 ... saw somebody attack a child.

Warm-up

Transparent people

A

CCTV[2] cameras

[2] **c**losed-**c**ircuit **t**ele**v**ision

B

telephone tapping

C

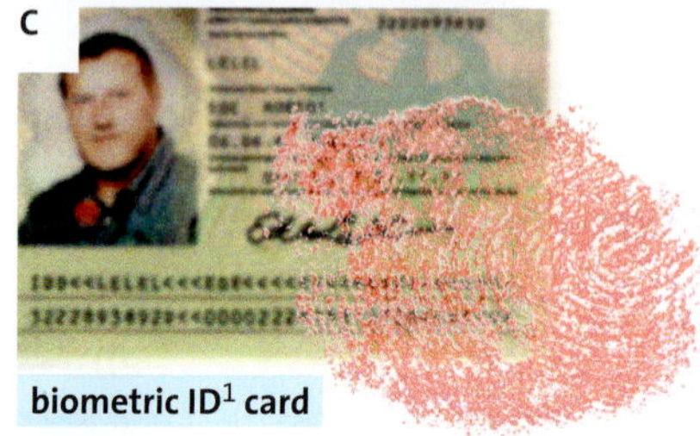

biometric ID[1] card

[1] **id**entification

D

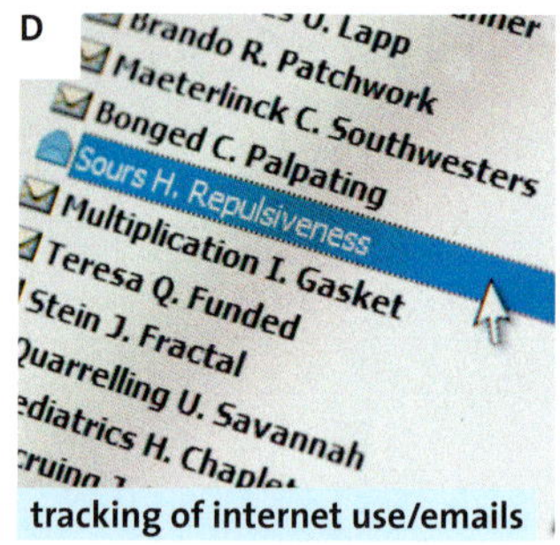

tracking of internet use/emails

E

GPS[3] tracking system

[3] **g**lobal **p**ositioning **s**ystem

F

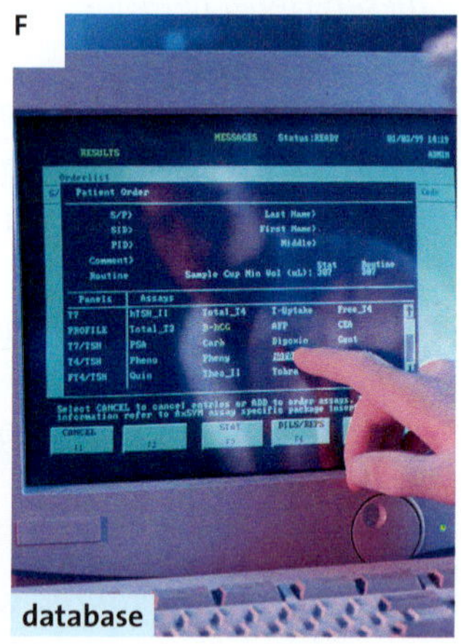

database

G

private security firms

H

informer phoning hotline

Which means of surveillance would you use to …

1. … find out somebody's true identity?
2. … find out whether somebody has been playing online games on the office computer?
3. … follow a getaway car after a bank robbery?
4. … guard a factory site at night?
5. … listen to people planning a crime on the phone?
6. … make complete crime statistics available to the police?
7. … report your neighbour for being cruel to his dog?
8. … see into dark corners in an underground car park?

Which means of surveillance do you find acceptable?

A Surveillance cameras

Person surveillance – street surveillance, store surveillance and infrastructure surveillance at, for example, airports and railway stations – is generally carried out with fixed or swivelling video cameras linked to a monitor and a digital video recorder. Increasingly, modern systems make use of wireless technology to avoid complicated cabling.

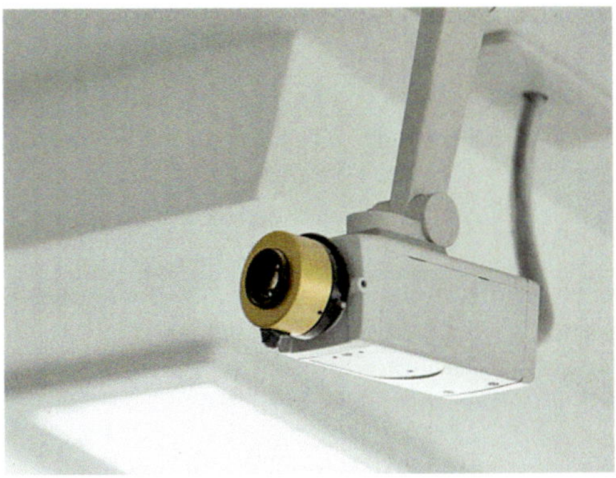

Most cameras are not hidden, but suspended from ceilings or mounted on poles. In fact, the idea is not really to catch wrongdoers, but to frighten them away. Often there is even a big notice telling people that they are under surveillance, like this one from a department store in the USA:

> **Don't even think of it!**
>
> **WARNING**
>
> **This store is under continuous video surveillance.**
> **This means that we are watching YOU all the time.**

Some systems, particularly those in exclusive locations like upmarket fashion boutiques and jewellers, high-class restaurants or the lobbies of expensive hotels, make use of more discreet "micro-cameras". These are often hidden in light fittings, say, so that customers do not feel spied on.

1 Tell somebody in German ...

1 ... where surveillance cameras are used.
2 ... why many modern systems use wireless technology.
3 ... why cameras are not hidden, but can be easily seen by customers.
4 ... why cameras are often hidden in more expensive shops, restaurants and hotels.

Technical specification sheet

Multiplex function:	Triplex (Record/Playback/Internet view)
Storage:	Two HDD bays, unlimited
Backup:	IP download or connect to VCR
Video:	Input, 4 channel; Output: 2 Ch (Monitor/VCA), 4 Ch loop VGA (optional) to computer monitor
Compression rate:	Basic: 12 KB; Normal: 16 KB; Enhanced: 20 KB; Fine: 24 KB; Super Fine: 28 KB
Recording/Search:	Motion Activated, Continuous
Recording speed:	Up to 120 fps
Languages:	English, Spanish, German, Italian, French
Live viewing screen:	1 or 4 screen, PIP, Freeze; 2x Digital Zoom; 120 fps
Network:	Ethernet (TCP), Remote Viewer software included
Power requirement:	DC 12V 10A power supply (included)
Weight:	Approx. 7 kg
Dimensions:	W 38 cm x L 28 cm x H 9 cm

2 **a) Link the abbreviations 1–10 to their meanings in the word snake. They are not necessarily in the same order.**

ampereschanneldirectcurrentframespersecondkilobytepictureinpicturetransmissioncontrolprotocolvideographicsarrayvoltagecontrolledamplifiervolts

1 KB	**3** PIP	**5** DC	**7** Ch	**9** fps
2 VCA	**4** TCP	**6** VGA	**8** A	**10** V

b) Now explain the abbreviations and their meanings in German.

3 Tr 30 **You are Cliona Price of Celtic Security Ltd. First listen to the telephone enquiry from John Fletcher of MPS Logistics. Note down the important information. Then write the text of an email recommending the CL-4MPEG4PACK Surveillance Package.**

Von: Cliona Price
An: John Fletcher
Betreff: CL-4MPG4PACK Surveillance Package

Dear Mr Fletcher
Further to our telephone conversation this morning, we are pleased to recommend the CL-4MPEG4PACK Surveillance Package to you.
This package comprises …

4 **John Fletcher of MPS Logistics likes the look of the surveillance package that Cliona Price recommends so he phones her to ask for a formal offer. Read Cliona's offer and then say if the statements are true, false or you can't say because the information is not given. Correct the false statements.**

CELTIC SECURITY SYSTEMS LIMITED

Unit 12–16, Jamestown Business Park • Dublin 11, Ireland
Phone +353-1-8654024-12 • Fax +353-1-8654024-15 • Email sales@css.ie

MPS Logistics Limited
30 Mill Road
Cork
County Cork

Your ref.: ---
Our ref.: CP/S/MPS/1
12 September 20..

Dear Mr Fletcher

Offer

Thank you for your telephone call of 11 September and we are pleased to submit[1] you the following offer:

CL-4MPEG4PACK Surveillance Package comprising four (4) Sony surveillance cameras, Sony Digital Video Recorder, DVR Security Surveillance, CCTV System, complete with all fixing devices / cabling **€1843.00**

As agreed on the phone, this price includes delivery to your Cork business premises[2] and installation by our technicians.
The price is 10% below list price for a first order. We also offer a further cash discount[3] of 2% for payment in full within 15 days of installation.
Delivery and installation will take place immediately on receipt of order[4].
Thank you again for your enquiry and please contact us if you have any questions.
We look forward to receiving an order from you and are sure that you will be satisfied in every way.

Yours sincerely
Celtic Security Systems Limited

Cliona Price

Cliona Price
Sales Department

1 *einreichen*
2 *Firmengelände*
3 *Skonto*
4 *Auftragseingang*

1 The letter is from MPS Logistics Ltd to Celtic Security Systems Ltd.
2 MPS Logistics' business premises are in Cork.
3 Cliona does not have a reference for MPS Logistics yet.
4 Cliona's reference is a standard reference that she always uses.
5 The letter is dated 14 September 20..
6 The offer is made in answer to an email.
7 The offer is for a surveillance package.
8 Celtic Security's price is much lower than that of other firms.
9 The buyer is responsible for installing the equipment.
10 The supplier offers a total discount of 12%.

5 **You work in the Buying Department of MPS Logistics Ltd in Cork. Submit an order for the surveillance package to Celtic Security Systems Ltd. Use the language and form of Cliona's offer as a model. Date your letter ten days after Cliona's and use your own name.**

6 **Work in small groups. First look at the plan of a filling station with shop. Where would you position video cameras so that the whole area (including all parts of the shop) is under continuous surveillance?**

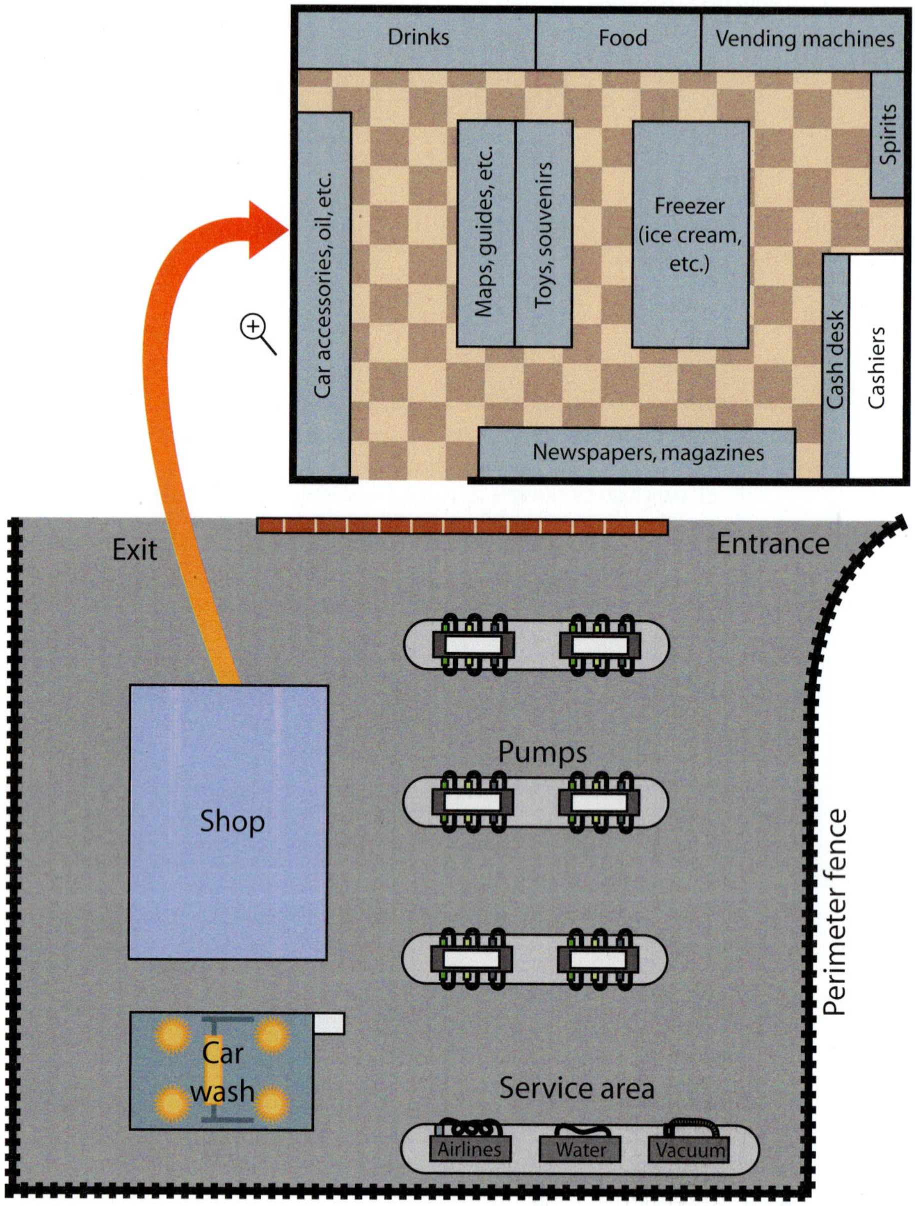

Here's a start:

Outside, I'd **position** cameras at the entrance and exit to the filling station. I'd **place** a further camera at the pumps so that car number plates are clearly visible. ...

In the shop, I'd **put** one camera (in the corner) between the drinks and the car accessories because then/so that the cashiers can see (what's happening) behind the maps and guides. I'd **locate** another camera ...

7 Listen to the radio discussion and choose the correct answer, a), b) or c).

Tr 31

1	The radio programme is called	**a)** "Stars tonight". **b)** "Star witness". **c)** "Expert witness".
2	The programme presenter's name is	**a)** Clare Morris. **b)** John Sutter **c)** James Sutton.
3	Clare is a	**a)** criminal lawyer specialising in civil rights. **b)** human rights lawyer specialising in privacy law. **c)** commercial lawyer specialising in business takeovers.
4	The discussion topic is	**a)** security and privacy. **b)** the use of video cameras. **c)** surveillance and terrorism.
5	Clare says digital communication has led to	**a)** people having little or no privacy. **b)** a big fall in street crime. **c)** much greater safety.
6	As an example of digital communication James gives	**a)** credit cards. **b)** GPS tracking. **c)** mobile phones.
7	Stored data can be accessed	**a)** only with great difficulty. **b)** by absolutely anybody. **c)** only by a person with the right skill and equipment.
8	The number of times you will be photographed in Oxford Street is almost	**a)** 360. **b)** 400. **c)** 410.

8 Writing

Which view do you agree with most? Why/Why not? Write a letter to the editor of the programme giving your own opinion.

I have nothing against video cameras and even accessing people's computers and so on. Why should I? I have nothing to hide and it makes me feel safer. I'm glad that somebody is watching over me. And online shopping's so convenient. I don't mind retailers knowing what I buy and what interests me. People shouldn't be so hysterical.

I don't like the feeling of being spied on – you know, "Big Brother is watching you." I use the internet, sure, but only as an encyclopedia or dictionary. I never just surf around or visit chat rooms, for example. And I never write anything in an email or text message that I wouldn't write on a postcard.

B Electronic article surveillance (EAS)

Opinion surveys about how we see different crimes put shoplifting near the bottom of the list. In fact, most people don't seem to think that shoplifting is really a crime at all. This tolerant attitude is surprising when you think that the cost of shoplifting – over 20 thousand million dollars a year in the USA alone – pushes up prices and makes shopping much less pleasant for the rest of us. Now read on to find out more.

Shopkeepers or "retailers" try to stop people shoplifting in three ways, or a mix of them.

Firstly, they can make it as difficult as possible to remove goods from the store. They do this by locking goods away, joining them together with wires, keeping them behind the counter or simply putting them out of reach on a high shelf. However, these measures, though effective, are unpopular among shoppers, who have to call a sales assistant before they can look at articles more closely or try on clothing, for example.

EAS labels

Secondly, retailers can watch everybody who comes into the store by using store detectives or CCTV cameras. Some even give generous discounts to honest customers who report shoplifters. But a lot of shoppers don't like being spied on, either, and tend to choose stores without such obvious surveillance methods.

This dislike of conventional anti-shoplifting methods among shoppers has led many retailers to adopt the third means of reducing shoplifting, i.e. electronic article surveillance (EAS) or "tagging". This highly effective system, which you will now find in practically all modern stores, has reduced shoplifting in clothing stores, for example, by an amazing 80%.

In EAS systems, a hard plastic tag or a label is fixed securely to every article in the store. When an article has been paid for, the tag is removed by a shop assistant with a special tool, or the label is deactivated by swiping it with a fixed or handheld electronic scanner. If the tag has not been removed or the label has not been deactivated, then an alarm will sound when the article is taken through the EAS gates at the entrance of the store. In addition, a photo may be taken of the potential shoplifter.

The most common EAS system is the radio frequency (RF) system that you will find out about here. This is the one that is most probably used in your local stores.

9 **Answer the questions.**

1. Why is it surprising that many people have such a tolerant attitude to shoplifting?
2. What do shoppers have against making it difficult to remove goods from the store?
3. Why do shoppers tend to go to stores that do not make use of store detectives and/or CCTV cameras?
4. What is the big difference between surveillance cameras and EAS?
5. How are EAS labels deactivated?
6. What happens if somebody tries to take a tagged article out of the store?

10 Tr 32

You are a technical adviser with a firm selling surveillance equipment to retailers. In the extracts below, you are talking to Jane Hudson, the owner of an exclusive fashion store. Complete the extracts with words and information from the drawing *and* your answers to Exercise 9. Then listen to the CD to check your answers and fill in any empty gaps.

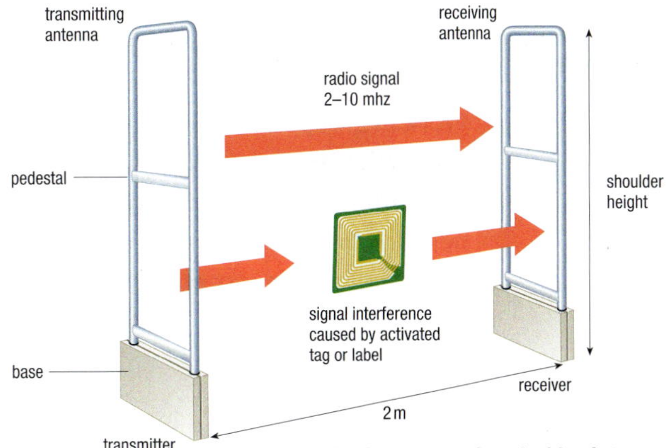

EAS exit gate seen from inside of store

1 **Jane** I understand that EAS is effective against ...[1], but how does it work?
You Well, the complete system is made up of two parts. First, there are two ...[2] that are positioned about two ...[3] apart just inside the entrance to the store. This is what we call an 'EAS ...[4]'.

2 **Jane** I see – and the second part?
You The second part of the system is hard ...[1] or electronic labels that are fixed securely to every ...[2] on sale in the shop.

3 **Jane** Fine, but what's the difference between a tag and a label?
You Well, hard plastic tags are removable with a special ...[1], so they can be reused many times. Labels, on the other hand, are just ...[2] by swiping them with a ...[3] so they leave the shop with the article.

4 **Jane** Okay, I've got that. Now how about this exit gate?
You Well, looking from the inside of the store, the left hand pedestal has a radio ...[1] in its base connected to an antenna which reaches to about an adult's ...[2]. This pedestal continuously transmits a preset radio ...[3] to the ...[4] of the right hand pedestal, which is in fact a radio ...[5].

5 **Jane** Fine. What happens then?
You Well, I don't want to get too technical, but the ...[1] of the radio signal at the exit gate can be set at any value between two and ...[2] megahertz. This means that the gate can handle a huge number of different tag and label ...[3].

6 **Jane** I think you've lost me. I'm afraid you'll have to explain that.
You No problem. The point is that the tags and the labels are all fitted with miniature ...[1] and ...[2] of a given frequency, too. When activated, they respond to the radio ...[3] across the exit gate.

7 **Jane** Really? Good heavens.
You Now, if a tag has **not** been removed or a label has **not** been deactivated, it ...[1] with the radio signal across the exit gate and an ...[2] sounds. In more advanced systems, a hidden ...[3] takes a photo of the person leaving the store as well.

11 Giving a talk

Work in groups of 3 or 4. You work for a consumer protection organisation. You have been asked to give a talk on mobile phone security. Read the introduction and then use the outline to prepare the talk, adding any ideas of your own. (The internet sites could help you here.) When you are satisfied with your talk, appoint a member of the group to give it before the rest of the class.

Introduction

Mobile phone theft is a big problem throughout the developed world. For example, the British Crime Survey says that there was an 8% rise in street theft during the first quarter of 2007. Just over 50% of the increase was the result of phone theft.

Mobile phones are protected by a PIN code on the SIM card, but this is easy to crack (*knacken*) either technically or by the use of violence. Worldwide IMEI (**I**nternational **M**obile station **E**quipment **I**dentity) numbers that give each phone a "unique identity" are not much more effective. This is because they are not stored in one central worldwide databank and, even if they were, they, too, are relatively easy to reprogram, thus changing the phone's identity.

For this reason, the industry is looking for more reliable ways of protecting mobile phones by making them useless to thieves. Go to the UK's Mobile Industry Crime Action Forum website at www.micaf.co.uk to find out more.

Outline: Effective ways of protecting mobile phones from theft

Introduction
Problem of phone theft as cause of street crime; find out crime statistics for your country/state/region; ...

Development I
Ineffectiveness of traditional means of protection (PIN codes, IMEI numbers); can be cracked technically; use of violence/threatening behaviour (*bedrohliches Verhalten*) to find out codes/numbers

Development II
Review more effective means of protection, e.g. Remote XT system, ...

Conclusion
Choose most effective means, giving brief reasons why

Useful websites
- Mobile Industry Crime Action Forum website at www.micaf.co.uk
- Get Safe Online at www.getsafeonline.org
- How Stuff Works at www.howstuffworks.com (protecting cellphones)

Practising grammar

Remember:
Verb + *to*-infinitive, verb + *-ing* form → Begleitgrammatik, Abschnitt 21, Seite 160

1 The boss **wants to install** surveillance cameras in the canteen.
2 The boy **admitted shoplifting** when the EAS alarm went off.
3 Did you **remember to send** MPS Logistics an offer, Cliona?
4 I'm sorry. I can't **remember meeting** you before.

Complete the text with the pairs of verbs. Be careful about some tenses.

For many years, the police *failed to stop* [1] (fail/stop) the rise in violent street crime in Hackney, one of the poorest parts of London. Whatever they did, crime just ... [2] (keep/rise). Things were so bad that local people ... [3] (avoid/use) the streets after dark and many ... [4] (stop/go out) at all. They ... [5] (prefer/stay) at home, where they felt safe. In fact, things became so bad on some big housing estates (*Wohnanlagen*) that people ... [6] (refuse/live) on them at all.

Luckily, today that ... [7] (seem/be) changing fast. Why?

About two years ago, local business people ... [8] (begin/work) with the police to make Hackney's streets safe again. They ... [9] (decide/install) CCTV cameras throughout the borough (*Stadtbezirk*). Surveillance experts also ... [10] (suggest/make) use of new software that could recognise the faces of the one hundred or so worst criminals, hooligans and thugs in the area.

As a first step, they ... [11] (agree/put up) cameras just along main roads. When it became clear that these cameras ... [12] (tend/keep) criminals off the streets, the police ... [13] (continue/install) cameras until almost the whole borough was covered.

Crime figures, at least those for street crime, immediately ... [14] (begin/fall) and they ... [15] (not stop/fall) ever since. For example, the CCTV cameras ... [16] (succeed in/reduce) car theft and theft from cars by nearly 40% last year.

When local people were recently asked if they ... [17] (dislike/live) "under the eye of a camera", over 90% said that they ... [18] (not mind/have) them in the streets at all. Many said that they ... [19] (enjoy/go) to the pubs and cafés again. In fact, many locals said they ... the police ... [20] (want/set up) even more cameras around Hackney.

Warm-up

1
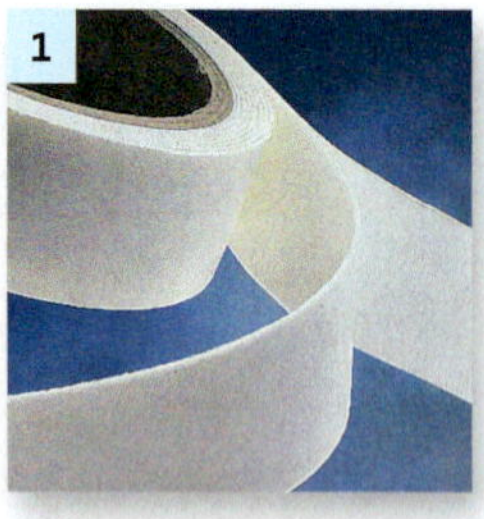
2
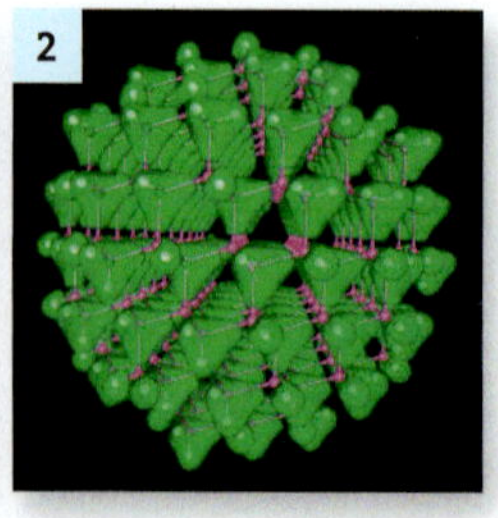
3
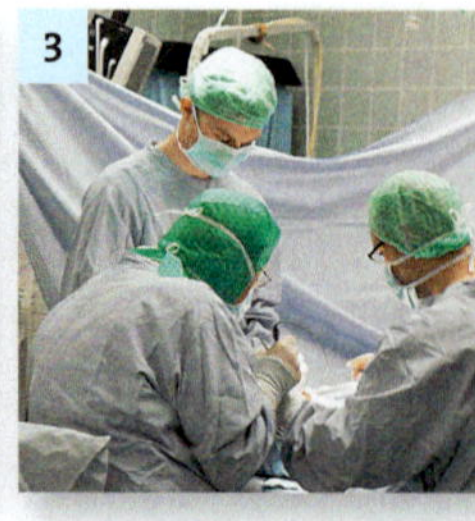
4
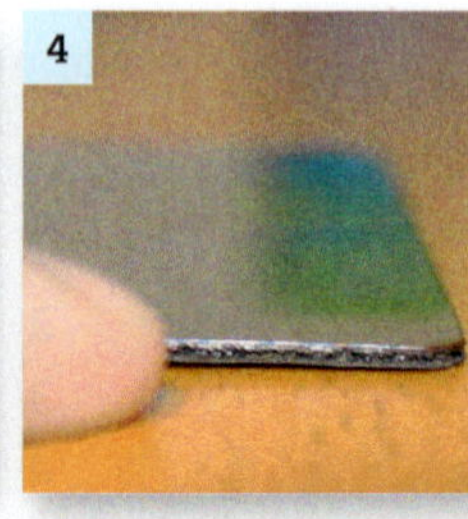
5

6

7

8

a **Better than Bulbs**

b NEW STEEL COPIES BIRDS' HOLLOW BONES

c Concrete Brings Light into Dark Corners

d **Nanotech Key to High-Power, Long-Life Batteries**

e **All GM Ingredients Must Be Labelled**

f NEW GLUE BONDS PLASTIC TO METAL

g **The World's Best Listener**

h **Nanoengine to Explore Human Body?**

Do the assignments.

- Link the photos 1–8 to the headlines a–h.
- Which of the photos and/or headlines do you find most positive for the future? Which do you find less interesting? Put them in your personal order of importance. Give reasons.

Brainstorming

Work in small groups. First brainstorm possible future technical developments and/or inventions. Then choose one particularly promising invention. Think of a specific application and point out your invention's advantages and disadvantages. Finally, compare your results with those of other groups.

A FUTURE LIGHTING TECHNOLOGY

Complete the description of the structure of an incandescent (Edison) light bulb with words and expressions from the drawing. Then listen to the CD to check your answers and fill in any empty gaps.

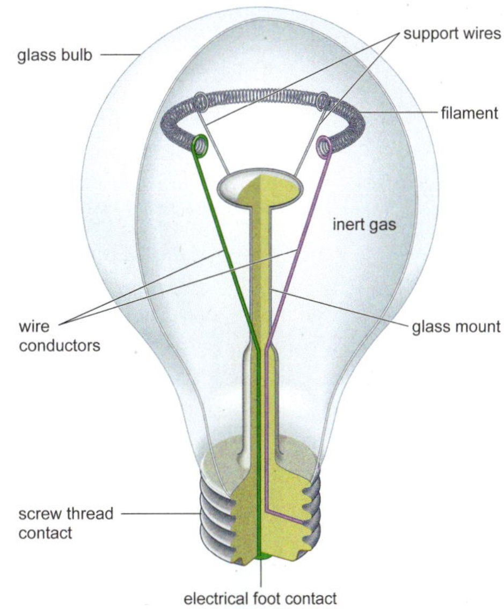

Tr 33 At its base, an incandescent light bulb has two metal contacts, an outer … …[1] contact and an inner … …[2] contact. The metal contacts connect with two … …[3], which carry current to and from the …[4]. The filament is held in place by two … …[5] joined to a vertical … …[6]. The whole arrangement (*Vorrichtung*) is housed in a sealed … …[7] filled with an … …[8], generally argon.

Now read on.

The big problem with Thomas Edison's incandescent light bulb is that it is extremely inefficient, converting only about 10% of the electricity it consumes into light. The remaining nearly 90% is transformed into excess heat which, in some devices, has to be dissipated by electrically-powered cooling fans. If that's not crazy, what is? For this reason, Australia and Ireland are already phasing out incandescent bulbs by law, and the USA is about to do so at an estimated saving of $18 billion* in household electricity bills.

Halogen light bulb

But are halogen light bulbs any better? Probably not. Although they are more durable and produce more light per unit of electricity than Edison's bulb does, they are also more expensive to buy and generate even more excess heat. Furthermore, at a temperature of around 500°C, this heat can cause injury by burning and even fires. That's why halogen bulbs almost always have a 12V power supply in domestic applications.

Low-energy compact fluorescent light bulbs (CFLs) are very much more efficient, durable and hence ecologically friendly than either incandescent or halogen bulbs. However, CFLs, too, are sometimes criticised for environmental reasons because of their use of highly toxic mercury. But this is unlikely to be a problem for long. As the magazine article on the next page suggests, the day of CFLs may be over before it has really begun.

Compact fluorescent light bulb

* AE: 1,000,000,000

Particularly in regions with few daylight hours during long winters, lighting consumes a huge amount of electricity, which has in turn often been generated in CO_2-intensive power plants. This has given rise to a sometimes angry debate about the pros and cons of different light bulbs, and even to the banning of less efficient incandescent and halogen bulbs in favour of low-energy CFLs in some countries. But, as this article from an American magazine shows, this debate may soon be over.

Accident may mean the death of the light bulb

It's almost certain that the main source of light in the future won't be Edison's famous light bulb or even a CFL, but a wall, a piece of furniture or a plate, say.

Sadly for makers of light bulbs, an accidental discovery suggests that LED lighting could soon be a cheaper, more durable and more ecologically-friendly alternative to both the conventional light bulb and to fluorescent lighting.

LEDs are already used in many applications, but they all have one thing in common. Although LEDs use less electricity than conventional lighting, the illuminated area tends to be very limited – a torch, say, or a mobile phone display or even Christmas tree candles.

But all that was before Michael Bowers, a graduate student at Vanderbilt University in Nashville, Tennessee, made a strange accidental discovery. Bowers was actually trying to make really small quantum dots. These are crystals which are normally rather small, less than one-thousandth of the width of a human hair, in fact.

However, quantum dots contain between 100 and 1,000 electrons. They are easily excited little bundles of energy, and the smaller they are, the more excited they become. Each quantum dot in Bowers' particular batch was exceptionally small, containing only 33 or 34 pairs of atoms.

One property of quantum dots is that if they are subjected to light or if an electric charge is applied to them, they generate light themselves, generally a bright, vivid colour. But when Bowers subjected his dots to a laser beam, a surprising and unexpected thing happened.

"I was surprised when a white glow covered the table," Bowers said. "The quantum dots were supposed to emit blue light, but instead they were giving off a beautiful white glow."

Then Bowers and another student had the idea of mixing the dots with polyurethane and coating a blue LED light bulb with the mixture. The lumpy bulb wasn't pretty, but it produced white light similar to a regular light bulb.

Bowers' device gives off a comfortably warm, yellowish-white light that shines twice as brightly and is 50 times more durable than a standard 60-watt Edison light bulb.

This was a real breakthrough because until quite recently LEDs could only produce green, red and yellow light, which limited their use.

Then blue LEDs were developed, which were later adapted to produce a whitish light with a light-blue hue. These LEDs were an improvement, but still not ideal.

Now, however, that has changed.

LEDs emit twice as much light as an ordinary 60-watt bulb and burn for over 50,000 hours. The Department of Energy estimates LED lighting could reduce U.S. energy consumption for lighting by 29 per cent by 2025. LEDs don't emit much heat, so they're also more energy-efficient. And they're much harder to break.

If the new quantum dot lighting goes into series production, light won't come just from light bulbs. Quantum dot mixtures could be painted on just about anything and electrically excited to produce a rainbow of colours, including white.

1 **Do the assignments.**

1 Try to name more applications of LEDs.
2 Pick out information about quantum dots.
3 Describe the reaction of quantum dots when exposed to light.
4 Explain why Michael Bowers was surprised when he subjected his batch of quantum dots to a laser beam.
5 Point out why Bowers' accidental discovery can be called a "breakthrough".
6 Say why the US Department of Energy is interested in LED technology.

2 **Work with a partner. Imagine that you are designing the lighting for a studio apartment with bathroom and kitchen. Draw a plan and show where you would put in lighting, and of what type.**

3 **a)** The verbs and nouns in the table are all used in the introduction or the report on pages 141–143. Try to provide the missing form or forms.

	Verb	Noun
1	develop	(2) * * *
2	convert	(2) * * *
3	consume	(2) * * *
4	produce	(2) * * *
5	* * *	supply
6	* * *	discovery

	Verb	Noun
7	suggest	* * *
8	* * *	application
9	contain	* * *
10	* * *	improvement
11	emit	* * *
12	reduce	* * *

b) Use six of the words you have found in sentences of your own. Then compare your sentences with those of a partner.

4 Tr 34

You are about to hear an extract from *Ask the experts*, a radio phone-in show that answers listeners' questions on home improvement. The show is chaired by Paul Hansen. Listen to the extract and do the assignments.

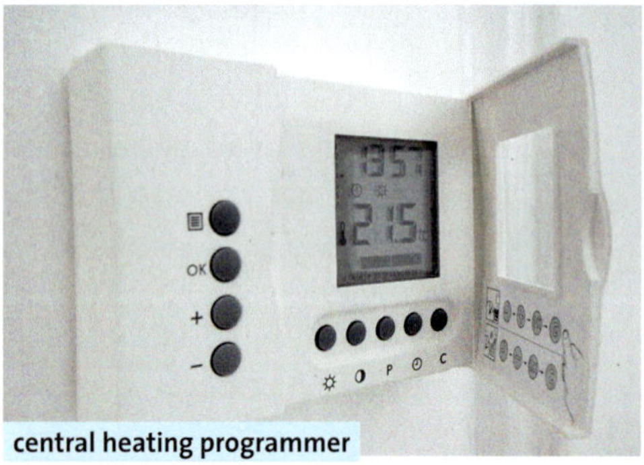
central heating programmer

radiator thermostat

heat insulation material

condensing boiler

a) Work with a partner. Choose the word or expression you hear, a) or b). They are in the same order and come in the first half of the dialogue.

1	**a)** heating our homes	**b)** heating our houses
2	**a)** fuel bill	**b)** oil bill
3	**a)** price increases	**b)** price rises
4	**a)** home heating systems	**b)** domestic heating systems
5	**a)** heating company	**b)** heating firm
6	**a)** warmer clothes	**b)** warmer clothing
7	**a)** insulation material	**b)** isolation material
8	**a)** technical fixes	**b)** technological fixes
9	**a)** houseowners	**b)** householders
10	**a)** much more effectively	**b)** much more efficiently
11	**a)** ten to twelve per cent	**b)** ten or eleven per cent
12	**a)** heat-reactive cameras	**b)** heat-sensitive cameras

b) Use numbers from 1 to 10 to put the points in the outline (*Gliederung*) **into the correct order.**

A Benefit of central heating programmer/radiator thermostat
B Brief introduction of guests
C Cost benefit of heat exchangers
D Harnessing ground heat with heat exchangers
E Importance of modern insulating materials
F Statement of tonight's topic
G True fuel saving through use of condensing boiler (Mark Davis)
H Ecological benefit of heat exchangers
I Use of condensing/non-condensing boilers (Lynn Wood)
J Words of welcome

5 **Work with a partner. One of you works for Mark Davis's heating company, the other is a customer who is thinking of insulating his or her home. Use the drawings to ask questions about insulation and to explain its benefits.**

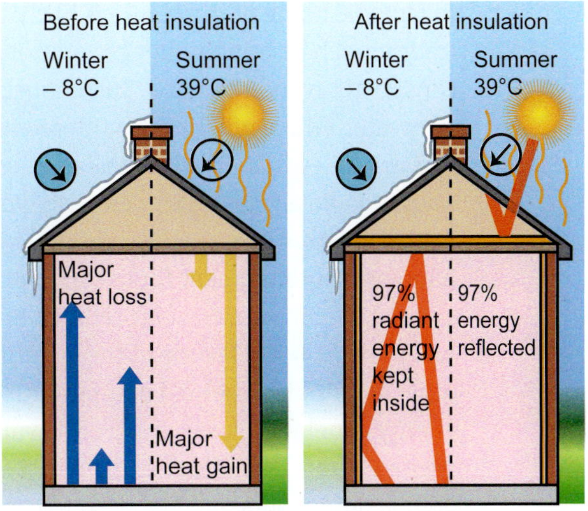

B THE HOME OF THE FUTURE

6 **First read the text and provide captions for the photos. Then use line numbers to suggest where they could be positioned in the text to assist understanding.**

a

b

c

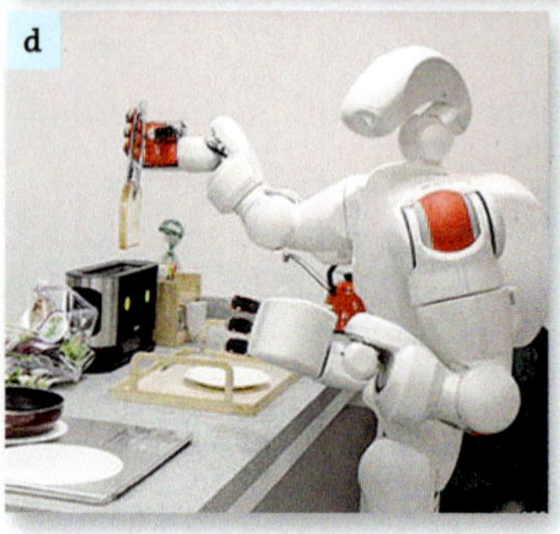

d

e

f

Until quite recently, the home of the future was seen as a fully-wired, high-tech wonder with every kind of communications device and labour-saving gadget, including robots to do the chores. But just a few years later, people are already losing interest in the so-called 'smart home'. Why? Because it depends upon huge amounts of electricity, and electricity generation is expensive, dirty and threatens the planet.

In fact, the home of the future is likely to be an updated, ecological version of traditional houses of earlier generations, with a strong emphasis on energy conservation and recycling. Such houses will have butts front and back to collect rainwater, which will then be filtered and used as so-called 'greywater' to flush toilets and maybe even to transport heat in heating radiators or underfloor tubes. Such houses could also be equipped with wood-fired stoves to reduce dependence on central heating systems. Increasingly, heating will be carried out by a heat exchanger harnessing ground heat for use in the house. To reduce dependency on electricity from generating companies, solar cells, voltaic roof tiles and a roof-mounted wind turbine will provide at least some of the electricity needed to power the heat exchanger and other essential household equipment. However, the most efficient and economical energy generation loses much of its value if significant amounts of energy are then wasted. Hence, the effective insulation of roofs and walls, triple-glazed windows and double outer doors must be used to conserve heat energy.

The house of the future

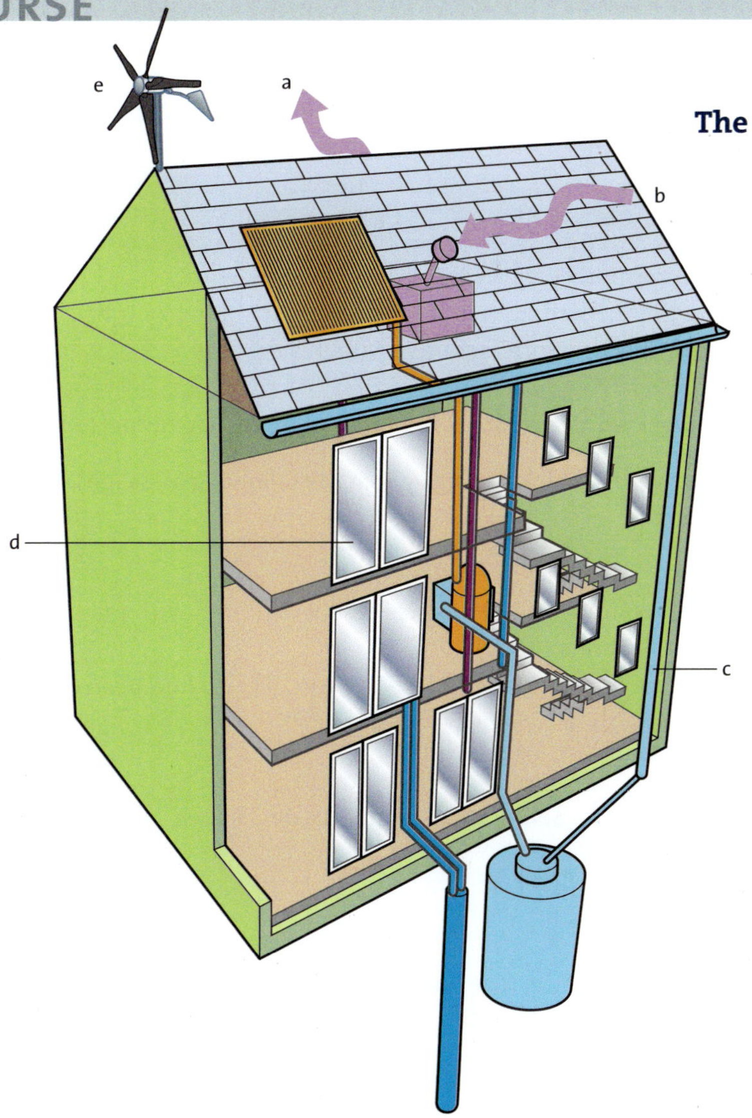

f ...

g ...

h ...

i ...

j ...

7 **Work in small groups.**

a) Match the letters *a* to *e* to the relevant items in the following box.

Air supply to heat exchanger • Exhaust air from heat exchanger • Rainwater recycling • Triple-glazed windows, airtight frames • Wind turbine

b) Now match the missing explanations *f* to *j* to the relevant items in the following box.

Ground source heat pump • Heat recovery ventilation system • Photovoltaic tiles • Solar thermal insulation • Total-wrap insulation of whole building

8 **Writing**

Write a description of the eco-house comparing it to a fully-wired smart house. Then present your paper to the rest of the class.

Practising grammar

Remember: Reported statements → Begleitgrammatik, Abschnitt 22, Seite 160

1 Björn **said** he **wanted** to travel to Ireland by ferry.
2 He **told Tara** that he **had missed** the ferry because of heavy traffic.
3 He **explained** that he **would arrive** in Dublin a day late.

1 **Tom Davis is a manager with Essex Solar Technology Ltd. You are his personal assistant.**

Use the introductory verb in brackets to give Tom the following messages. The first one has been done for you.

1 Anton Moser:
"My flight from Vienna has been cancelled but I will call with my new arrival time later, okay?" (phone to say)
*Anton Moser **phoned to say** (that) **his** flight from Vienna **had been cancelled** but **he would call** with **his** new arrival time later.*

2 Ted:
"We have a big order for solar panels from Austria. That will help us a lot."
(say)
Ted said (that) we ...

3 Mary:
"I may be a little late for the meeting because I have a dental appointment."
(tell me)
Mary told me that she might ...

4 Jill Snow:
"It's nearly the end of the month and I still haven't had Tom's expenses for last month. I must have them today or we will have problems."
(call to tell you; say)
Jill Snow called to tell you that it was ... She said she had to ...

5 Marcel:
"I won't be able to visit your stand at the trade fair (*Messe*) next month because I have a new job in the company."
(say)
Marcel ...

6 Sam (printer):
"I'm sorry, your new catalogue won't be ready before the end of next week because our supplier delivered the wrong paper."
(tell me)
Sam ...

Remember: Reported questions → Begleitgrammatik, Abschnitt 22, Seite 160

1 The boss **asked if** the report **was** ready yet.
2 Ben **wanted to know where** the new components **were**.

2 **Put the questions into reported speech. Use *asked* or *wanted to know* as introductory verbs.**

1 Ben: "Why are the machines not running?"
2 Joan: "Have you already sent the email, Terry?"
3 Brian: "What has happened to the printer?"
4 Barbara: "Have you seen the drawings anywhere, Tom?"
5 John: "Can you help me, please, Ann?"
6 Alice: "Did you find the keys to the cupboard, Ben?"
7 Joe: "Have you finished that report on the Linz contract yet, Brian?"
8 Ann: "What are you doing at the weekend, Mary?"
9 Tom: "How many spare bulbs do we have, Julie?"
10 Lisa: "Did you remember to order some more paper, Simon?"

Remember:
Reported instructions and requests → Begleitgrammatik, Abschnitt 22, Seite 160

1 The boss **told us to repair** the automatic lathe first.
2 Mary **asked Brian not to arrive** late for the meeting.

3 **Greg Brown, a safety officer, is talking to some new trainees. Report what he said to them with *told us*.**

1 "Don't go into the factory without a hard hat and ear muffs."
2 "Always use the yellow walkways."
3 "Watch out for forklift trucks near swing doors."
4 "Don't speak to machine operators when they're working."
5 "Keep away from the assembly lines."
6 "Don't wear jewellery or loose clothing."
7 "Make sure you know and follow the safety rules."
8 "Don't take food or drink into the factory. Use the canteen."
9 "Never smoke in the factory. Always go to a special smoking area."
10 "Don't run, and keep your eyes open at all times."

Anmerkung

Die Grammatik, die in den Schaukästchen exemplarisch dargestellt ist, wird hier in der Reihenfolge ihres Erscheinens behandelt.

Eine Ausnahme ist der **Refresher Course**, der einen getrennten Vorkurs bildet. Im **Refresher Course** ist die Grammatik – sehen Sie die Übersicht unten – an Ort und Stelle erklärt.

Übersicht des Refresher Course – Grammatik

Grammatik in den Schaukästchen (Main Course, Units 1–10)

1 Das *simple present*

→ Unit 1, Ex 1, 2, 3

1 We **supply** security systems.
2 Our technicians **work** all over Europe.
3 Björn **reads** the job ads every week.

- Mit Ausnahme der 3. Person Singular *(he, she, it)* hat das ***simple present*** dieselbe Form wie der Infinitiv (1, 2). Die **3. Person Singular** endet auf **-*s*** (3).
- Wir gebrauchen das ***simple present*** für permanente Zustände (1, 2) und sich ständig wiederholende Ereignisse bzw. Vorgänge (3).
- Daher steht das ***simple present*** häufig mit den folgenden **Zeitangaben**:
  - always, often, sometimes (*manchmal*), seldom (*selten*), never, ...
  - generally (*im Allgemeinen*), normally, usually (*gewöhnlich*), ...
  - every day/week/month/year/..., every morning/afternoon/...
  - on Mondays/Tuesdays/...

2 Das *present progressive*

→ Unit 1, Ex 4, 5

1 I **am reading** a good book **at the moment**.
2 Björn **is looking** for a job in Ireland **now**.
3 We **are advertising** for technicians **at present**.
4 The students normally have English on Mondays but today they **are having** French.

- Das ***present progressive*** bilden wir mit ***am*/*is*/*are* + *-ing*-Form**.
- Wir benutzen das ***present progressive*** für Vorgänge, die im Moment des Sprechens oder Schreibens tatsächlich passieren (1, 2, 3). Es wird auch für vorübergehende und außergewöhnliche Situationen benutzt (4).
- Das ***present progressive*** kommt oft mit den folgenden **Zeitangaben** vor:
  – at the moment (*im Augenblick*), now, at present (*zurzeit*)

Beachten Sie, dass die folgenden Verben normalerweise keine ***present progressive***-Form haben:

be, believe (*glauben*), hear, know (*wissen, kennen*), like, mean (*bedeuten*), notice (*bemerken*), see, seem (*scheinen*), want, wish (*wünschen*)

3 Kurzantworten

→ Unit 1, Ex 6

- **Kurzantworten** kommen im Englischen viel häufiger als im Deutschen vor. Sie werden benutzt, um ja/nein-Antworten (*yes/no answers*) weniger abrupt klingen zu lassen.
- Um Kurzantworten zu bilden, verwenden wir das **Verb** aus dem **Fragesatz** mit dem entsprechenden **Pronomen**.
- Bei **bejahten** Kurzantworten benutzen wir ausschließlich die **Langform**. Bei **verneinten** Antworten dagegen wird die **Kurzform** bevorzugt.

Ja/Nein-Frage	Kurzantwort (+)	Kurzantwort (–)
Is this your book?	Yes, **it is**.	No, **it isn't**./No, **it's not**.
Are you German?	Yes, **I am**.	No, **I'm not**
Are they here today?	Yes, **they are**.	No, **they aren't**./No, **they're not**.
Has Olga got the rule?	Yes, **she has**.	No, **she hasn't**.
Have you got a chisel?	Yes, **I have**.	No, **I haven't**.
Can you speak German?	Yes, **I can**.	No, **I can't**.
Is there a vice on the bench?	Yes, **there is**.	No, **there isn't**.

4 Frageanhängsel

→ Unit 1, Ex 7

- **Frageanhängsel** (deutsch: *nicht wahr?, gell?, oder?* usw.) verwendet man, wenn man sich vergewissern will.

- Man wiederholt das Hilfsverb des Aussagesatzes und fügt ein Personalpronomen hinzu, z.B.:

Björn is going to Ireland, **isn't he**? You haven't met him, **have you**?

- Wenn der Satz kein Hilfsverb enthält, verwendet man ***do/does/did***, z.B.:
  Tara works for CSS, **doesn't she**? Björn didn't apply by post, **did he**?

- Merke, dass das Frageanhängsel verneint ist, wenn der Aussagesatz bejaht ist, und umgekehrt.

5 Das *simple past*

→ Unit 2, Ex 1, 2, 5

1 Björn **reached** Emmerich at about 6 o'clock.
2 He **had** coffee at Gouda.
3 He **drove** off the motorway into a car park.
4 Björn **didn't reach** Rotterdam until 8.40.
5 **Did** Björn **have** breakfast at Gouda?
6 Why **did** he **decide** to take a later ferry?

- Bei **regelmäßigen Verben** bilden wir das ***simple past*** durch das Anhängen von ***-ed*** an den Infinitiv. Diese Form gilt bei allen Personen (1).

- Die **unregelmäßigen Verben** haben eine **Sonderform**, die man sich einfach merken muss (2, 3).

- Ist kein Hilfsverb vorhanden, verneinen wir mit ***didn't/did not*** (4). Fragen werden ebenfalls mit ***did*** gebildet (5, 6).

- Wir benutzen das ***simple past***, um **völlig abgeschlossene** Handlungen bzw. Situationen in der Vergangenheit zu schildern.

- Häufig wird das ***simple past*** mit folgenden **Zeitangaben** benutzt:

  - yesterday (*gestern*), the day before yesterday (*vorgestern*)
  - last night/month/summer/June/Christmas/...
  - two/three/... minutes/hours/weeks/years ago
  - in 1994/...

6 Das *past progressive*

→ Unit 2, Ex 3, 4, 5

1 At 7.40 Björn **was driving** towards the Gouda exit.
2 As Björn **was parking** his car some lorry drivers **were getting** ready to leave.
3 Björn **was driving** near Rotterdam when he **saw** another tailback.
4 The cars **were not/weren't moving** very fast.
5 **Were** the lorry drivers **having** breakfast?

- Wir bilden das ***past progressive*** mit ***was/were* + *-ing*-Form** des Verbs.
- Da bei dem ***past progressive*** immer ein Hilfsverb (***was/were***) vorhanden ist, werden die **Verneinung** mit ***not/n't*** (4) und **Fragen** durch **Umstellung** (5) gebildet.
- Möchten wir ausdrücken, dass eine Handlung zu einem bestimmten Zeitpunkt (1) oder während eines bestimmten Zeitraumes (2) in der Vergangenheit im Gange war, benutzen wir das ***past progressive***.
- Wir benutzen das ***past progressive*** auch, um zu verdeutlichen, dass eine Handlung im Gange war, als ein neues Ereignis (plötzlich) eintrat (3).

7 Das *will*-Futur

→ Unit 3, Ex 1

1 I**'ll**/I **will tell** Sally that you have arrived.
2 I'm sure you **will** soon **get to know** everybody.
3 It **won't/will not be** easy to find something good.
4 **Will Björn and Anton take** the flat in Dublin?

- Bei allen Personen bilden wir das ***will*-Futur** mit ***will* + Infinitiv** des Vollverbs (1, 2).
- Die **Verneinung** wird durch das Einfügen von ***n't/not*** unmittelbar nach ***will*** gebildet. Beachten Sie dabei die Kurzform ***won't*** (3).
- Um **Fragen** zu bilden, tauschen wir ***will*** und das Subjekt einfach aus (4).
- In erster Linie benutzen wir das ***will*-Futur** für zukünftige Situationen oder Ereignisse, die mit Sicherheit eintreten werden (1), für Vermutungen und Hoffnungen (2) und für Vorhersagen (3, 4).

8 Das *present perfect*

→ Unit 3, Ex 2, 3, 4

1 Björn **has been** in Dublin **for two days**.
2 The other trainees **have** already **met**.
3 Anton **hasn't/has not found** a flat yet.
4 **Have you thought** of sharing a flat, Björn?

■ Wir bilden das ***present perfect*** mit ***have/has*** **+ 3. Form** des Vollverbs (1, 2).

■ Die **Verneinung** wird durch das Einfügen von ***n't/not*** unmittelbar nach ***have/has*** gebildet (3).

■ Um **Fragen** zu bilden, tauschen wir ***have/has*** und das Subjekt einfach aus (4).

■ Möchten wir verdeutlichen, dass ein Zustand bzw. eine Handlung irgendwann in der Vergangenheit begonnen hat und noch andauert, benutzen wir das ***present perfect***, oft mit ***for*** bzw. ***since*** (1).

■ Wenn wir ausdrücken wollen, dass irgendetwas geschehen ist, ohne den Zeitpunkt des Ereignisses zu nennen, benutzen wir ebenfalls das ***present perfect*** (2, 3, 4).

■ Häufig benutzen wir das ***present perfect*** mit folgenden Zeitangaben:

- already (*schon*), still (*noch*)/still not, yet (*bisher*)/not yet (*jetzt noch nicht*)
- just (*gerade*), ever (*je(mals)*), not ever (*noch nie*), never
- for/since (*seit*)

9 *Some*, *any* und ihre Zusammensetzungen

→ Unit 4, Ex 1, 2

1 **Some lorry drivers** were having breakfast.
2 The hotel didn't have **any rooms**.
3 A Has Anton had **any luck** finding a flat?
B No, he hasn't found **anything** yet.
4 Anton worked **somewhere** in Graz.

■ ***Some*** und ***any*** werden unterschiedlich eingesetzt.

■ Wir benutzen ***some*** in **bejahten Aussagen** (1) und ***any*** in **verneinten Aussagen** (2) und **Fragen** (3).

■ Wir benutzen die **Zusammensetzungen** von ***some*** und ***any*** – ***somebody/anybody***, ***someone/anyone***, ***something/anything***, ***somewhere/anywhere*** – in genau derselben Art und Weise (3, 4).

10 Zählbare and nichtzählbare Substantiven

→ Unit 4, Ex 3

1 CSS is looking for **technicians** with good **qualifications**.
2 You must be in good **health** and have some **experience** of surveillance technology.
3 What **kinds of experience** have you got, Mr Drexler?
4 Please buy **a bottle of mineral water**.
5 I always carry **a can of petrol** in the car.

■ Substantive wie *technicians* und *qualifications* sind **zählbar**. Sie haben eine Pluralform und können mit *a/an* und Zahlwörtern benutzt werden (*a technician, two qualifications*) (1).

■ Substantive wie *health* und *experience* sind **nichtzählbar**. Sie haben keine Pluralform und können deswegen nicht mit *a/an* oder Zahlwörtern benutzt werden (2).

■ Wollen wir ein **nichtzählbares Substantiv** im Plural benutzen, dann müssen wir eine geeignete zählbare Mengenangabe – z. B. *kinds of* (3), *a bottle of* (4) oder *a can of* (5) – hinzufügen.

11 *Much, many, a lot of*

→ Unit 4, Ex 4

1 There are not **many 'pure materials'**.
2 Materials are not **much use** unless we can fix them together.
3 How **many synthetic materials** can you see?
4 Do computers use **much electricity**?
5 CSS exports **a lot of equipment** to Germany.
6 A chair must carry **a lot of weight**.

■ Wir benutzen ***many*** mit **zählbaren** und ***much*** mit **nichtzählbaren** Nomen in **verneinten Aussagen** (1, 2) und **Fragen** (3, 4).

■ In **bejahten Aussagen** benutzen wir ***a lot of*** mit **zählbaren** (5) sowie **nichtzählbaren** Nomen (6).

12 Relativpronomen

→ Unit 5, Ex 1

1 Markus is the trainee **who/that** comes from Hanover.
2 The firm makes equipment **which/that** can start jet engines.
3 Anton is the man **whose** family lives in Graz.
4 That is the computer **whose** disk drive doesn't work.

■ Ein **Relativsatz** ist eine Art **Nebensatz**, der durch ein **Relativpronomen** – ***who/that***, ***which/that***, ***whose*** – eingeleitet wird.

■ Wir benutzen **Relativsätze**, um den Hauptsatz durch zusätzliche Informationen genauer zu bestimmen.

■ Für **Personen** benutzen wir das Relativpronomen ***who*** bzw. ***that*** (1) und für **Sachen** ***which*** bzw. ***that*** (2).

■ Um **Besitz** bzw. **Zugehörigkeit** anzuzeigen, benutzen wir ***whose*** unmittelbar vor dem Substantiv bei **Personen** (3) und **Sachen** (4).

! **Beachten Sie**, dass Relativpronomen **immer dieselbe Form** haben.

13 Die Modalverben

→ Unit 5, Ex 2, 3, 4, 6

1 Tony **must arrange** a meeting about the Leipzig project.
2 Tony **mustn't forget** to telephone Bernd tomorrow.
3 Anton **needn't come** to the meeting.
4 Björn **can speak** English.
5 The trainees **can go** home now.
6 Lynn, **can** you **copy** the documentation, please?
7 The Cork project **may be** Björn's first big job for CSS.
8 **May** I **use** the telephone now?
9 Tony **should discuss** the agenda with Len Sullivan.

- Wir benutzen ***must***, um einen **Zwang** oder eine **dringende Notwendigkeit** auszudrücken (1).
- Das Modalverb ***must not/mustn't*** benutzen wir ausschließlich, um **Verbote** auszudrücken. Es entspricht dabei dem deutschen Modalverb „nicht dürfen" (2).
- Möchten wir sagen, dass es **keinen Zwang** oder **keine Notwendigkeit** gibt, dann benutzen wir ***need not/needn't*** (3).

Beachten Sie also, dass die **Verneinung** von ***must needn't*** – **nicht** ***mustn't*** – ist.

- Genau wie bei dem deutschen Modalverb „können" benutzen wir ***can***, um **Fähigkeit** (4), **Erlaubnis** (5) und **Bitten** (6) auszudrücken.

Beachten Sie, dass anders als im Deutschen ***can*** **nie alleine** steht:

- Wir benutzen ***may***, um **Möglichkeit** (7) oder **Erlaubnis** (8) auszudrücken.
- Das Modalverb ***should*** entspricht dem deutschen „sollte(n)". Wir benutzen es, um eine **dringende Empfehlung** auszudrücken (9).

14 Das *going to*-Futur, das *present progressive* mit zukünftiger Bedeutung

→ Unit 5, Ex 5

1 May and Alice **are going to give** a progress report.
2 Greg **is going to tell** us about quality assurance.
3 Switch off the machine. Can't you see that it **is going to overheat**?
4 We **are dealing** with the rest of the agenda **tomorrow**.
5 **After lunch tomorrow** Sally Peters **is coming** to the meeting.

- Das ***going to*-Futur** bilden wir mit ***am/is/are*** + ***going to*** + **Infinitiv** des Vollverbs (1, 2).

Beachten Sie, dass ***going to*** **und der Infinitiv** bei allen Personen unverändert bleiben.

- Wir benutzen das ***going to*-Futur** für zukünftige Pläne und Absichten (1, 2).
- Wir verwenden das ***going to*-Futur** auch für Ereignisse, die bald stattfinden werden. Dies ist dann der Fall, wenn bereits im Augenblick des Sprechens eindeutige Anzeichen dafür gegeben sind (3).
- Sehr oft benutzen wir das ***present progressive*** **und ein Adverb der zukünftigen Zeit**, um über schon entschiedene Pläne und Verabredungen zu sprechen (4, 5).

15 Das Passiv

→ Unit 6, Ex 1, 2

1. Equipment **is installed** immediately.
2. Some access roads **are** still **being built**.
3. This **was agreed** a year ago.
4. Work **has been delayed by bad weather**.
5. The first pumps **will be fitted** in May.

- Wir bilden das **Passiv** mit einer Zeitform von ***be*** und der **3. Form** des Vollverbs:

Zeit	Zeitform von be	3. Form
simple present	is/are	installed
present progressive	is/are being	built
simple past	was/were	agreed
present perfect	has/have been	delayed
***will*-future**	will be	fitted

- Ist der Verursacher einer Handlung **unbekannt** oder **zweitrangig**, benutzen wir das **Passiv**. In erster Linie steht also das **Ergebnis** des Vorganges im Vordergrund (1–3).
- Wollen wir den **Verursacher** eines Vorganges doch angeben, benutzen wir einen ***by*-agent** (4).

Beachten Sie dabei, dass der Agent mit ***by*** – auf keinen Fall mit ***from*** – gebildet wird.

16 Das Passiv mit Modalverben

→ Unit 6, Ex 3

1. Installation **can be started** at the end of May.
2. The equipment **must be stored** in Oxford.
3. The first systems **should be installed** in February.

■ Häufig gebrauchen wir Passivsätze, die mit **Modalverben** gebildet werden. Merken Sie sich vor allem die Anwendung des Musters **Modalverb + *be* + 3. Form** des Vollverbs. Dieses Muster bleibt **unverändert** bei allen Personen.

■ Hier sind weitere Beispiele:

Subjekt	Modalverb	be	3. Form	andere Teile
I	**can**	**be**	**reached**	at my hotel after 7.
The user's manual	**must**	**be**	**translated**	into German.
We	**may**	**be**	**told**	to attend.
All deliveries	**should**	**be**	**sent**	immediately.

17 Das Adverb

→ Unit 7, Ex 1

1 Sales fell **sharply** in the first half of the year.
2 Björn gives in too **easily** in an argument.
3 CNC lathes operate **automatically**.
4 This work schedule is **completely unrealistic**.
5 This chisel is **terribly blunt** (*stumpf*).
6 Don't go so **fast**, Tony. I haven't finished yet.
7 Björn is getting on **well** at CSS.
8 Electric motors convert energy **most efficiently**.
9 Robots can work **harder** and **longer** than human workers.

■ In erster Linie benutzen wir **Adverbien**, um **Verben** (z.B. 1, 2, 3) und **Adjektive** (4, 5) näher zu beschreiben.

■ Wir bilden die **meisten Adverbien** durch das einfache Anhängen von ***-ly*** an das entsprechende **Adjektiv** (1–5).

■ Bei dem Anhängen von **-ly** gibt es einige wichtige **Schreibregeln**.
Endet das Adjektiv auf

– **-y**, wird das **-y** zu **-i**:	eas**y**	→	eas**i**ly (2)
– **-le**, fällt das **-e** weg:	terribl**e**	→	terrib**ly** (5)
– **-ic**, wird **-ally** angehängt:	automat**ic**	→	automatic**ally** (3)

■ Einige Adverbien haben **dieselbe Form** wie das Adjektiv (6).
Die wichtigsten Adverbien dieser Gruppe sind:
– close (*nah*), early, fast, hard, high, late.

Beachten Sie: Die Adverbform von ***good*** ist ***well*** (7).

Die Steigerung von Adverbien

■ Endet das Adverb auf -*ly*, steigern wir mit ***more*/*most*** (8).

■ Hat **das Adverb dieselbe Form wie das Adjektiv**, steigern wir mit ***-er*/*-est*** (9).

Beachten Sie: Diese Steigerungen sind **unregelmäßig**:

– **well**	**better**	**best**
– **badly**	**worse**	**worst**

18 Die Steigerung von Adjektiven

→ Unit 7, Ex 2, 3

1. The **quickest** route is through the tunnel. It's **quicker** than the ferry.
2. The Calais – Dover route is **easier** than Hoek van Holland – Harwich.
3. Is The Hoek van Holland – Harwich route **more expensive** than Dunkerque – Dover?
4. I think Calais – Dover is the **best** route. It's **better than** Dunkerque – Dover.
5. The Hoek van Holland – Harwich route is **not as expensive as** the Channel Tunnel
6. Is the Dunkerque – Dover route **as long as** Calais – Dover?

- Einsilbige Adjektive (1) und zweisilbige Adjektive, die auf ***-y*** enden (2), werden mit ***-(i)er/-(i)est*** gesteigert.
- Mehrsilbige Adjektive werden mit ***more/most*** gesteigert (3).

Beachten Sie: **good – better – best**
bad – worse – worst

- Wollen wir Personen oder Sachen im Satz miteinander vergleichen, haben wir die folgenden Möglichkeiten:

– Unterschied	(*... als ...*)	**... than ...**	(4)
– Unterschied	(*nicht so ... wie ...*)	**not as ... as ...**	(5)
– Kein Unterschied	(*genauso ... wie ...*)	**as ... as ...**	(6)

19 *If*-Sätze, Typ I und II

→ Unit 8, Ex 1, 3, 4

1. If you **burn** petrol, you **produce** toxic emissions.
2. If we **do** nothing about air pollution, the situation **will become** worse.
3. If we **increased** petrol prices, drivers **would change** to public transport.

- Ein ***if*-Satz** besteht aus zwei Teilen: dem ***if*-Teil** und dem **Hauptteil**. Der ***if*-Teil** drückt eine **Bedingung** aus, der **Hauptteil** eine **Folge**.
- Bei ***if*-Sätzen** von **Typ I** und **Typ II** kommen folgende **Zeitenmuster** am häufigsten vor:

Typ I	*If*	+	simple present	+	simple present	(1)
	If	+	simple present	+	*will*-future	(2)
Typ II	*If*	+	simple past	+	*would* + Infinitiv	(3)

Beachten Sie, dass *will* und *would* **nie** im ***if*-Teil** stehen.

- Die **Typen von *if*-Sätzen** unterscheiden sich gemäß der **Wahrscheinlichkeit** der zu erwartenden Folge:

Typ I Die Folge ist **sicher** (1) bzw. **sehr wahrscheinlich** (2).
Typ II Die Folge ist **theoretisch möglich**, aber **weniger wahrscheinlich** (3).

20 *If*-Sätze, Typ III

→ Unit 8, Ex 2, 3

1 If Björn **had taken** a later ferry, he **would have arrived** in Hoek van Holland on time.
2 If he **had not stopped** for coffee, Björn **would have caught** the ferry.
3 **Would Björn have been** late if he **had left** the motorway in Gouda?

- Das **Satzmuster** von **Typ III-*if*-Sätzen** lautet:
  – *If* + past perfect + *would have* + 3. Form

- Da wir das Ergebnis schon kennen, ist die Folge der Bedingung im ***if*-Teil unmöglich** (1–3).

21 Verb + *to*-Infinitiv bzw. *-ing*-Form

→ Unit 9

- Der ***to*-Infinitiv** steht nach bestimmten Verben, z. B.:

agree, arrange, choose, decide, expect, fail, hope, learn, offer, promise, seem, refuse, tend, try, want, would like

- Nach einer Reihe weiterer Verben (teilweise mit Präposition) steht die ***-ing*-Form**, z. B.:

admit, agree with, apologize for, avoid, believe in, dislike, enjoy, imagine, keep, look forward to, mind, miss, practise, risk, stop, suggest, succeed in, talk about, worry about

- Nach einigen Verben kann man entweder die ***-ing*-Form** oder den ***to*-Infinitiv** verwenden, ohne dass sich die Bedeutung dabei ändert, z. B.:

begin, continue, hate, like, love, prefer, start

22 Indirekte Rede

→ Unit 10, Ex 1, 2, 3

Mit indirekter Rede (*reported speech*) kann man berichten, was jemand gesagt hat. Vgl: Direkte Rede: "My job **is** boring." > Indirekte Rede: She said her job **was** boring.

Wenn man direkte Rede in indirekter Rede wiedergibt, muss man einiges verändern.

Zeitformen

■ Wenn das einleitende Verb in der Vergangenheit (*said, asked, explained*, etc.) steht, muss man häufig die Zeitform nach folgendem Muster verändern:

present	**past**
"I speak German."	She said she spoke German.
past / present perfect / past perfect	**>past perfect**
"I went to London."	She said she had been to London.
"I've often played tennis."	She said she had often played tennis.
"I had never met him before."	She said she had never met him before."
***will*-future**	**> *would* + Infinitiv**
"I'll help them."	She said she would help them.
can	could
may	might
must	had to

Pronomen

■ Genau wie im Deutschen muss man bei einem Bericht oft, aber nicht immer, Pronomen verändern.

Direkte Rede	**Indirekte Rede**
"**My** friends help **me** and **my** children."	She said **her** friends helped **her** and **her** children.
"**They** lent **me their** car on Tuesday."	She said **they** had lent **her their** car on Tuesday.
"**I**'m going to get **you** some coffee."	She said **she** was going to get **me** some coffee.

■ Mit der indirekten Rede kann man auch Fragen und Befehle wiedergeben.

Fragen mit Fragewörtern

Direkte Rede	**Indirekte Rede**
How old are you?	She asked me how old I was.
Where do you live?	She asked where I lived.

Fragen ohne Fragewörter

Direkte Rede	**Indirekte Rede**
Do you like jazz?	She wanted to know **if** I liked jazz.
Can you play tennis?	She wondered **if** I could play tennis.

Befehle

Direkte Rede	**Indirekte Rede**
Stand up, please.	He told me **to** stand up.
Be quiet!	She ordered them **to** be quiet.
Don't talk so fast.	He told him **not to** talk so fast.

IRREGULAR VERBS

be	was/were	been	*sein*
become	became	become	*werden*
begin	began	begun	*anfangen*
break	broke	broken	*brechen*
bring	brought	brought	*bringen*
build	built	built	*bauen*
burn	burnt/burned	burnt/burned	*brennen*
buy	bought	bought	*kaufen*
catch	caught	caught	*fangen*
come	came	come	*kommen*
cost	cost	cost	*kosten*
do	did	done	*tun,* machen, erledigen
draw	drew	drawn	*zeichnen*
dream	dreamt	dreamt	*träumen*
drink	drank	drunk	*trinken,* saufen
drive	drove	driven	*fahren*
eat	ate	eaten	*essen,* fressen
feel	felt	felt	*sich* fühlen
find	found	found	*finden*
fly	flew	flown	*fliegen*
forget	forgot	forgotten	*vergessen*
go	went	gone	*gehen*
get	got	got	*bekommen,* erhalten
grow	grew	grown	*wachsen*
have	had	had	*haben*
hear	heard	heard	*hören*
hide	hid	hidden	*verstecken*
hit	hit	hit	*schlagen,* aufprallen auf
hold	held	held	*halten*
hurt	hurt	hurt	*verletzen*
keep	kept	kept	*behalten*
know	knew	known	*kennen,* wissen
lay	laid	laid	*legen*
lead	led	led	*führen*
learn	learnt	learnt	*lernen*
leave	left	left	*verlassen*
let	let	let	*lassen*
lose	lost	lost	*verlieren*
make	made	made	*machen*
mean	meant	meant	*bedeuten*
meet	met	met	*sich* treffen
pay	paid	paid	*(be-)zahlen*
put	put	put	*setzen,* stellen, legen
read	read	read	*lesen*
ring	rang	rung	*läuten,* klingeln, anrufen
rise	rose	risen	*steigen*
run	ran	run	*laufen*
say	said	said	*sagen*
see	saw	seen	*sehen*
sell	sold	sold	*verkaufen*
send	sent	sent	*senden,* schicken
shoot	shot	shot	*(er-)schießen*
show	showed	shown	*zeigen*
sing	sang	sung	*singen*
sit	sat	sat	*sitzen*
speak	spoke	spoken	*sprechen*
spend	spent	spent	*ausgeben,* verbringen (Zeit)
stand	stood	stood	*stehen*
take	took	taken	*nehmen*
teach	taught	taught	*unterrichten,* lehren
tell	told	told	*erzählen,* mitteilen, sagen
think	thought	thought	*denken,* meinen
understand	understood	understood	*verstehen*
wake	woke	woken	*wecken*
write	wrote	written	*schreiben*

NUMBERS AND MATHEMATICAL TERMS

Vermerk

Achten Sie in den nachfolgenden Darstellungen auf die Rechtschreibung der **fettgedruckten** Zahlen.

1 Kardinalzahlen

cardinal numbers

1	one	13	**thirteen**	25	twenty-five
2	two	14	fourteen	26	twenty-six
3	three	15	**fifteen**	27	twenty-seven
4	four	16	sixteen	28	twenty-eight
5	five	17	seventeen	29	twenty-nine
6	six	18	**eighteen**	30	**thirty**
7	seven	19	nineteen	40	**forty**
8	**eight**	20	twenty	50	**fifty**
9	nine	21	twenty-one	60	sixty
10	ten	22	twenty-two	70	seventy
11	eleven	23	twenty-three	80	**eighty**
12	twelve	24	twenty-four	90	ninety

100	one hundred
101	one hundred and one
200	two hundred
210	two hundred and ten
225	two hundred and twenty-five
1000	one thousand
1010	one thousand and ten
2000	two thousand
2225	two thousand two hundred and twenty-five
3578	three thousand five hundred and seventy-eight
10,000	ten thousand
10,381	ten thousand, three hundred and eighty-one
100,000	one hundred thousand
100,456	one hundred thousand, four hundred and fifty-six
1,000,000	one million
1,250,000	one million, two hundred and fifty thousand
2,146,251	two million, one hundred and forty-six thousand, two hundred and fifty-one

2 Ordinalzahlen
ordinal numbers

1st	first	13th	thirteenth	30th	thirtieth
2nd	second	14th	fourteenth	40th	fortieth
3rd	third	15th	fifteenth	50th	fiftieth
4th	fourth	16th	sixteenth	60th	sixtieth
5th	**fifth**	17th	seventeenth	70th	seventieth
6th	sixth	18th	eighteenth	80th	eightieth
7th	seventh	19th	nineteenth	90th	ninetieth
8th	**eighth**	20th	**twentieth**	100th	one hundredth
9th	**ninth**	21st	twenty-first		
10th	tenth	22nd	twenty-second		
11th	eleventh	23rd	twenty-third		
12th	**twelfth**	24th	twenty-fourth		

3 Mathematische Ausdrücke
mathematical terms

+	plus
–	minus
±	plus or minus
×	multiplied by
÷	divided by
=	is equal to/equals
≠	is not equal to/does not equal
≅	is approximately equal to
<	is less than
>	is greater/more than

%	per cent
‰	per thousand
$\sqrt{\ }$	root/square root
$\sqrt[3]{\ }$	cube root
4^2	four squared
5^3	five cubed
2^4	two to the power of four/two to the fourth
p, π	pi

Language focus

This firm has **twenty-two** service technicians.	**NICHT** „twenty and two"
We have three hundred **and** fifty employees.	**NICHT** „three hundred fifty"
This machine cost $275,000.	**NICHT** $275.000
This machine is two point **nought** five metres long.	(UK)
This machine is two point **zero** five metres long.	(US)
The temperature must be below **zero**.	
The code for the UK is double **oh**, double four.	
My account number is three-seven-**oh**-five-**oh**.	
Check the oil **once** a week.	**NICHT** „one time"
That machine broke down **twice** yesterday.	**NICHT** „two times"
Barry was late **three times** last week.	

TIME AND DATES

1 Uhrzeit: Die 12-stündige Analoguhr

time: the 12-hour analogue clock

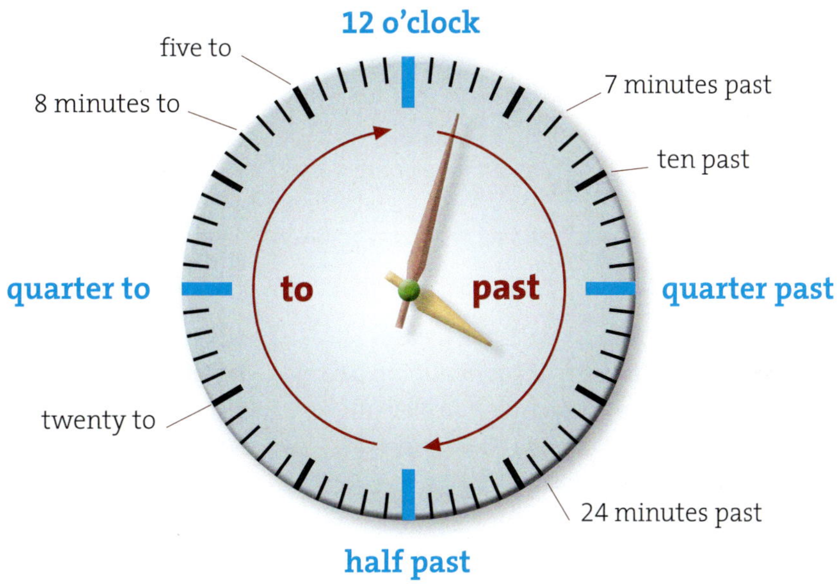

- Das Wort *o'clock* (= of the clock) wird nur bei vollen Stunden benutzt:
  It's five *o'clock*.
  Es ist fünf Uhr.

Beachten Sie, wie man die *Zeitangabe der halben Stunde* im Englischen ausdrückt:

It's *half past ten.*	It's *half past eleven.*
Es ist halb elf.	*Es ist halb zwölf.*

- Bei den *fünfminütigen-Schritten* können wir das Wort *minutes* weglassen:
  It's *twenty to four.*
  Es ist zwanzig vor vier.

- Bei den *Zwischenzeiten* müssen wir jedoch *minutes* hinzufügen:
  Our train goes at *seven minutes to eleven.*
  Unser Zug fährt um sieben vor elf.

- Die *Abkürzungen a.m. (= ante meridiem, vormittags) und p.m.* *(= post meridiem, nachmittags)* können nicht bei Zeitangaben mit *o'clock, past, half* und *to* benutzt werden:

The bus goes at *five a.m.*	Let's meet at *seven forty-five p.m.*
Der Bus fährt um fünf Uhr.	*Laß uns um neunzehn Uhr fünfundvierzig treffen.*

2 Uhrzeit: Die 24-stündige Digitaluhr

time: the 24-hour digital clock

- Bei Digitaluhren liest man die Zahlen einfach vor:
  It's *nine thirty-five*.
  Es ist neun Uhr fünfunddreißig.
  It's *eleven thirteen*.
  Es ist elf Uhr dreizehn.

- Bei vollen Stunden nach 12 Uhr mittags benutzt man das zwölfstündige System:
  It's *seven* (*o'clock*).
  Es ist neunzehn Uhr.
  It's *twelve* (*midnight*).
  Es ist vierundzwanzig Uhr.

Language focus

What time is it?	**NICHT** „How late is it?"
What's the time?	
Can you *tell* me the time, please?	**NICHT** „Can you say me ...?"
The delivery arrived at *about* ten o'clock.	**NICHT** „against ten o'clock"
The meeting is from ten *to* eleven.	**NICHT** „from ten until/till eleven"
The meeting starts *at half past two*.	**NICHT** „at half two"
The fax arrived *at two minutes to four*.	**NICHT** „at two to four"
The plane lands *at ten a.m.*	**NICHT** „at ten o'clock a.m."
I'll see you *at seven p.m.*	**NICHT** „at nineteen o'clock"
It *is* Wendy's birthday tomorrow.	**NICHT** „Wendy has birthday tomorrow"

3 Das Datum

dates

Days of the week		months		seasons
Monday	Mon	January	Jan	spring
Tuesday	Tues	February	Feb	summer
Wednesday	Wed(s)	March	Mar	autumn
Thursday	Thurs	April	Apr	winter
Friday	Fri	May	–	
Saturday	Sat	June	Jun	
Sunday	Sun	July	Jul	
		August	Aug	
		September	Sept	
		October	Oct	
		November	Nov	
		December	Dec	

- Merken Sie sich vorweg, dass wir die Namen von Wochentagen und Monaten immer groß schreiben:
  Today is *Friday*, 30 *August*.
  Jahreszeiten dagegen werden klein geschrieben:
  We bought the new machine last *autumn*.

- Bei der *Angabe von Daten* unterscheiden wir zwischen der *Schreibweise* und der *Aussprache*.

Wir schreiben:	**Wir sagen:**
on 6 June, on 6th June	on *the sixth of* June
on June 6th	on June *the* sixth (UK), on June sixth (US)

Beachten Sie: Bei der *Schreibweise* können wir entweder Ordinalzahlen oder Kardinalzahlen benutzen. Bei der *Aussprache* dagegen werden nur Ordinalzahlen verwendet.
In der Industrie werden Kardinalzahlen – z. B. *22 April 1997* – bevorzugt.

- Vermeiden Sie auf alle Fälle Datumsangaben, die nur aus Zahlen bestehen, z. B. *6.11.09* oder *6/11/09*. In Europa bedeutet diese Angabe *6 November 2009*, in den USA und in einigen asiatischen Länden – u. a. Japan – jedoch *11 June 2009*.

Beachten Sie auch die Aussprache von Jahresangaben:

1900	nineteen hundred
1906	nineteen-oh-six
1989	nineteen eighty-nine
2000	two thousand
2009	two thousand (and) nine
2010	twenty ten

DECIMALS AND FRACTIONS

1 Dezimalzahlen

0.2	nought *point* two	(UK)
	zero *point* two	(US)
0.035	nought *point* nought three five	(UK)
	zero *point* zero three five	(US)
2.75	two *point* seven five	
12.80	twelve *point* eight	
267.23	two hundred and sixty-seven *point* two three	
7.3'	seven *point* three *recurring*	
	sieben Komma drei unendlich	

■ *Beachten Sie unbedingt*, dass im Englischen Dezimalzahlen mit einem *Punkt* (.) geschrieben werden. Dies ist auf Grund der Schreibweise von Zahlen größer als 999 wichtig, da hierzu ein Komma verwendet wird, z.B. 2,569,348 (Two million, five hundred and sixty-nine thousand, three hundred and forty-eight).

2 Brüche

1/4	a/one quarter
1/3	a/one third
1/2	a/one half
2/3	two thirds
3/4	three quarters
2/5	two fifths
1/6	a/one sixth
3/8	three eighths
1/100	a/one hundredth
3/1000	three thousandths
2 1/4	two *and* a quarter
4 1/2	four *and* a half
5 3/10	five *and* three tenths

Language focus

The temperature is thirty-seven *point* five.	**NICHT** „thirty-seven comma five“
The process lasts one *and* a quarter hours.	**NICHT** „one a quarter“

UNITS OF MEASURE

Vermerk

In der Wirtschaft und Wissenschaft benutzt Großbritannien das metrische System (*metric system*). Darüber hinaus werden das *pint* und die *mile* aus Kostengründen weiter verwendet.

In anderen Teilen der englischsprachigen Welt werden die herkömmlichen britischen Maßeinheiten immer noch benutzt. *Beachten Sie* dabei, dass bei Hohlmaßen (*capacity*) und Gewichten (*weight*) die USA zum Teil ihre eigenen US-Einheiten haben. Eine Übersicht der UK/US-Maßeinheiten finden Sie auf dem vorderen Innenumschlag.

Language focus

Can you get a pint *of* paint, please?	**NICHT** „pint paint"
My old car needed two litre*s* of oil. (UK)	**NICHT** „two litre"
Twelve inches make a foot and three *feet* a yard.	**NICHT** „three foot(s)"

Umrechnungstabelle

Conversion table

from	into	multiply by
pounds	grams	453.6
pounds	kilograms	0.4536
US ton	tonnes	0.907
inches	centimetres	2.54
feet	metres	0.3048
yards	metres	0.9144
miles	kilometres	1.609
sq inches	sq centimetres	6.452
sq feet	sq metres	0.0929
acres	hectares	0.405
sq mile	sq kilometres	2.59
UK pints	litres	0.568
US pints	litres	0.473
UK gallons	litres	4.546
US gallons	litres	3.785

Basic word list

Diese Liste enthält ca. 500 Grundwörter, die in **Technical English at Work – The new edition** als bekannt vorausgesetzt werden. Nicht angeführt, jedoch vorausgesetzt sind Strukturwörter und einige sehr elementare Wörter wie einfache Präpositionen, Pronomen, Zahlen sowie genaue Entsprechungen wie z. B. *hotel, restaurant*.

about *über, etwa*
above *über, oben*
actually *tatsächlich*
add *zusammenzählen, hinzufügen*
after *nach*
afternoon *Nachmittag*
again *wieder*
age *Zeitalter, Alter*
ago *vor*
airport *Flughafen*
all *alle(s)*
almost *fast, beinahe*
alone *allein(e)*
already *schon, bereits*
also *auch, außerdem*
although *obwohl*
always *immer*
another *noch eine*
answer *Antwort, Lösung; (be)antworten*
any *irgendetwas, -welche, jede*
anybody *jemand, jede/r*
anyone *jemand, jede/r*
anything *etwas, alles*
anywhere *irgendwo(hin)*
apple *Apfel*
around *herum*
arrive *ankommen*
as *wie, als, da*
ask *fragen, bitten*
away *weg, entfernt*
back *zurück*
bad *schlecht, schlimm*
bag *Tasche*
bath *Bad*
because *weil*
become *werden*
bed *Bett*
beer *Bier*
before *vor(her)*
begin *anfangen, beginnen*
behind *hinter, hinten*
believe *glauben*
below *unter, unten*
best *beste/r/s, am besten*
better *besser*
big *groß*
bit, a ~ *ein bisschen, etwas*
black *schwarz*
blue *blau*
body *Körper*
book *Buch; buchen, bestellen*
both *beide*
boy *Junge*
breakfast *Frühstück*
bring *bringen, holen*
brother *Bruder*
but *aber, sondern*
buy *kaufen*
bye *tschüs*
can *dürfen, können*
car *Auto*
card *Karte*
carry *tragen*
chair *Stuhl*
change *(Ver-)Änderung; (aus)wechseln, (sich) ändern*
cheap *billig, günstig*
child *Kind*
choose *(aus)wählen*
classroom *Klasse(nzimmer)*
clock *Uhr*
coffee *Kaffee*
cold *kalt*
come *kommen, geliefert werden*
conversation *Gespräch, Unterhaltung*
cost *Kosten; kosten*
country *Land, Staat*
cup *Tasse*
cut *schneiden*
dark *Dunkelheit; dunkel*
date *Datum, Termin*
day *Tag*
different *anders, verschieden*
difficult *schwer, schwierig*
do *tun, machen*
dog *Hund*
door *Tür*
down *unten*
draw *ziehen, zeichnen*
drink *Getränk; trinken*
drive *(Auto) fahren*
driver *Fahrer/in*
e.g. *z. B.*
each *jede/r/s*
early *früh*
east *Osten*
easy *leicht, einfach*
eat *essen*
egg *Ei*
either *entweder*
electric *elektrisch, Strom-*
electronic *elektronisch*
else *andere/r/s*
empty *leer*
end *Ende, Schluss; (be)enden*
enough *ausreichend, genug*
etc. *usw.*
even *sogar (noch)*
evening *Abend*
ever *je(mals)*
every *jede/r/s*
everybody *jede/r*
everything *alles*
everywhere *überall*
example *Beispiel*
excellent *ausgezeichnet, hervorragend*
excuse me *Entschuldigung, Verzeihung*
expensive *teuer*
fact *Tatsache*
fall *fallen; Fall; Herbst*
false *falsch*
family *Familie*
far *weit (entfernt)*
fast *schnell*
father *Vater*
feel *(sich) fühlen*
few *ein paar, wenig/e*
final *letzte*
finally *schließlich, endlich*
find *finden, suchen*
fine *gut, schön, fein*
finish *Ende, (be)enden*
fire *Feuer; entlassen*
first *erste/r/s, zuerst*
flat *flach; Wohnung*
follow *(be)folgen*
food *Essen, Nahrung*
foot *Fuß*
forget *vergessen*
free *kostenlos, frei*
Friday *Freitag*
friend *Freund/in*
front *Vorderseite*
fuel *Kraft-, Treibstoff*
full *voll*
full-time *Vollzeit-*
further *weitere*
future *Zukunft; (zu)künftig*
garden *Garten*
gas *Gas, Benzin*
get *holen, bekommen, werden*
girl *Mädchen*
give *geben*
go *gehen, fahren*
good *gut*
good evening *guten Abend*
good morning *guten Morgen*
good night *gute Nacht*
goodbye *auf Wiedersehen*
government *Regierung*
great *groß(artig)*
half *Hälfte; halb*
happy *glücklich, froh, zufrieden*
hard *schwer*
hardly *kaum*
head *Kopf*
health *Gesundheit*

hear *hören*
help *Hilfe; helfen*
here *hier*
hide *verstecken, verbergen*
high *hoch; äußerst*
home *Zuhause, Heim; nach Hause*
hope *hoffen*
hot *heiß, warm*
hour *Stunde*
house *Haus*
how *wie*
how many *wie viele*
how much *wie viel*
human *menschlich*
idea *Idee, Gedanke*
if *wenn, falls, ob*
ill *krank*
I'm afraid *leider*
important *wichtig*
information *Auskunft, Information(en)*
inside *innerhalb, drinnen*
instead *stattdessen, anstatt*
interested *interessiert*
interesting *interessant*
international *international*
interview *Interview, Vorstellungsgespräch*
interviewer *Interviewer/in*
into *in ... hinein*
introduce *(sich) vorstellen, (miteinander) bekannt machen, einführen*
January *Januar*
jet *Düse(nflugzeug)*
job *Arbeit, Stelle*
July *Juli*
just *einfach, nur, genau*
keep *(be)halten*
key *Schlüssel, Taste*
kind *nett; Art, Sorte*
kitchen *Küche*
know *kennen, wissen*
knowledge *Wissen, Kenntnis(se)*
language *Sprache*
large *groß, umfangreich*
last *letzte/r/s, zuletzt*
late *spät*
learn *lernen, erfahren*
leave *(ver)lassen*
left *linke/r/s, links; übrig*
leg *Bein*
less *weniger, abzüglich*
lesson *Lektion, Unterricht(sstunde)*
let *erlauben, (zu)lassen, vermieten*
let's *lasst uns*
letter *Brief, Buchstabe*
life *Leben*
light *Licht, Leuchte; hell, leicht*
like *(ähnlich) wie; mögen, gern tun*
likely *wahrscheinlich*
list *Liste; auflisten, notieren*
listen *zuhören*
little *klein, wenig*
live *wohnen, leben*
local *örtlich, lokal*
long *lang*
look *Blick; (aus)sehen, blicken*
lot, a ~ *viel, sehr*
lots of *viel, viele*
loud *laut*
love *Liebe; lieben, sehr gern mögen*
low *niedrig*
lunch *Mittagessen*
machine *Maschine*
main *Haupt-, wichtigste*
mainly *hauptsächlich*
make *machen*
man *Mann, Mensch*
manager *Manager/in*
many *viele*
map *Karte*
March *März*
market *Markt*
may *dürfen, können, mögen*
maybe *vielleicht*
mean *bedeuten, meinen, heißen*
meet *(zusammen)treffen, begegnen*
meeting *Konferenz, Besprechung*
memo *Memo, (Akten-)Notiz*
metal *Metall*
method *Methode, Verfahren*
metre *Meter*
middle *Mitte*
mind *Geist; etw dagegen haben*
minute *Minute*
moment *Moment*
Monday *Montag*
money *Geld*
month *Monat*
more *mehr*
morning *Morgen*
mother *Mutter*
mouth *Mund*
much *viel*
museum *Museum*
must *müssen*
name *Name; (be)nennen*
nationality *Staatsangehörigkeit*
nature *Natur*
near *nahe*
nearly *beinahe, fast*
necessary *nötig, erforderlich*
need *Bedarf; brauchen, benötigen*
negative *negative, verneinend*
never *nie(mals)*
new *neu*
news *Neuigkeit(en), Nachricht(en)*
newspaper *Zeitung*
next *nächst; danach*
nice *schön, nett*
night *Nacht*
nobody *niemand*
normally *normal(erweise)*
north *Norden*
note *Notiz; beachten, notieren*
nothing *nichts*
now *nun, jetzt*
nowadays *heutzutage*
number *Nummer, Zahl*
o'clock *... Uhr*
office *Büro, Amt*
often *oft, häufig*
old *alt*
once *einmal, einst, sobald*
only *nur, einzig*
open *öffnen, beginnen*
opinion *Meinung, Einschätzung*
opposite *Gegenteil; gegensätzlich*
orange *orange*
other *andere/r/s*
out *aus*
over *(vor)über*
own *eigene; besitzen*
page *Seite*
pair *Paar*
paper *Papier, Zeitung*
part *Teil, Rolle*
partner *Partner/in*
past *vorbei, nach; Vergangenheit*
pay *Gehalt; (be)zahlen*
people *Personen*
per cent *Prozent*
perhaps *vielleicht, eventuell*
person *Mensch, Person*
pet *Haustier*
phone *Telefon; anrufen*
photo(graph) *Foto*
phrase *(Rede-)Wendung, Ausdruck*
picture *Bild*
piece *Stück, Teil*
pilot *Pilot/in*
place *Stelle, Platz; setzen, stellen*
plan *Plan; planen*
play *spielen*
please *bitte*
pleased *vergnügt*
plenty *viel, reichlich*
poor *arm, schlecht, mangelhaft*
position *Stellung*
positive *positiv, bejahend*
possible *möglich, denkbar*
post *Post, Pfahl*
power *Kraft, Strom; antreiben*
practice *Übung*
practise *(ein)üben*
prefer *vorziehen, bevorzugen*
press *drücken*
private *privat, persönlich*
probably *wahrscheinlich*
product *Produkt, Erzeugnis*
production *Produktion, Herstellung*
programme *Programm*
project *Projekt*
push *schieben*
put *setzen, stellen, legen*
qualification *Qualifikation*
question *Frage*
quick *schnell, rasch*
quiet *still, ruhig*
quite *ziemlich, ganz*
rather *ziemlich, lieber*
read *lesen*
ready *fertig, bereit*
real *echt, wirklich*

really *wirklich, eigentlich, tatsächlich*
reason *Vernunft, Grund*
remember *sich erinnern, daran denken*
rest *Rast, Rest*
right *rechts, rechte/r/s, richtig; Recht*
river *Fluss*
road *(Land-)Straße*
rock *Felsen, Stein*
room *Zimmer, Raum*
run *laufen (lassen)*
safe *Safe; sicher*
same *gleiche/r/s, der-, die-, dasselbe*
Saturday *Sonnabend, Samstag*
saw *sägen*
say *sagen*
school *Schule*
sea *See*
seat *Sitz(platz)*
second *zweite/r/s; Sekunde*
secretary *Sekretär/in*
see *sehen*
seem *(er)scheinen*
sell *(sich) verkaufen*
send *senden, schicken*
sentence *Satz*
set *Satz; setzen*
several *etliche, einige, mehrere*
shirt *Hemd*
shop *Laden, Geschäft; einkaufen*
short *kurz, klein, knapp*
should *solle/n, sollte/n*
show *zeigen*
side *Seite, Rand*
simple *einfach*
since *da, weil, seit*
sister *Schwester*
sit *sitzen, sich hinsetzen*
situation *Situation, Lage*
sleep *Schlaf; schlafen*
slow *langsam; verlangsamen*
small *klein*
so *also, damit, deshalb, so (dass)*
social *sozial*
soft *weich, leise, sanft*
some *einige, etwas*
somebody *jemand*
someone *jemand*
something *etwas*
sometimes *manchmal*
somewhere *irgendwo(hin)*
soon *bald*
sorry *traurig; Verzeihung*
sort *Sorte, Art; sortieren*
south *Süden*
space *(Zwischen-)Raum, Abstand, Platz, Weltall*
speak *sprechen, reden*
spell *buchstabieren, schreiben*
stairs *Treppe(n)*
stand *Stand; stehen, aushalten*
standard *Norm, Standard, Maßstab; normal, üblich*
start *Beginn; anfangen, starten, beginnen*
stay *Aufenthalt; bleiben*
still *still, trotzdem, (immer) noch*
stop *Halt; (an) halten, aufhören (mit)*
straight *gerade*
street *Straße*
strong *stark, kräftig*
student *Student/in, Lernende/r*
study *lernen, studieren, untersuchen, betrachten*
such *so, solch*
summer *Sommer*
sun *Sonne*
Sunday *Sonntag*
supermarket *Supermarkt*
sure *freilich, sicher(lich)*
table *Tisch, Tabelle*
take *nehmen, bringen, dauern*
talk *Gespräch, Unterhaltung, Vortrag ; sprechen, reden*
teach *unterrichten, lehren*
teacher *Lehrer/in*
technology *Technik, Technologie*
telephone *Telefon*
television *Fernsehen, Fernseher*
tell *sagen, erzählen*
than *als*
thank *danken*
thank you *danke*
thanks *Dank; danke*
that *das, dass*
then *dann*
there *da, dort(hin)*
there are *es sind*
there is *es gibt*
thick *dick*
thin *dünn*
thing *Sache, Ding, Gegenstand*
think *denken, meinen, finden, glauben*
this *dies, diese/r/s*
through *durch*
Thursday *Donnerstag*
ticket *Karte, Fahrschein*
time *Zeit, Mal*
tip *Hinweis, Tipp, Vorschlag*
today *heute*
together *zusammen*
toilet *Toilette, WC*
tomorrow *morgen*
tonight *heute Abend/Nacht*
too *zu, auch*
tooth *Zahn*
top *Spitze, Gipfel; Spitzen-*
total *(End-)Summe; völlig, (ins) gesamt*
tour *Tour, Rundreise*
towards *auf ... zu*
town *Stadt*
train *Zug*
training *Training, Übung*
travel *Reisen; reisen, fahren*
tree *Baum*
true *richtig, wahr*
try *versuchen, probieren*
Tuesday *Dienstag*
twice *zweimal*
type *Typ; erfassen, tippen*
under *unter*
understand *verstehen, begreifen*
until *bis*
use *benutzen, verwenden*
useful *nützlich*
user *Anwender/in, Benutzer/in*
usually *normalerweise, gewöhnlich, meistens*
very *sehr*
village *Dorf*
visit *Besuch; besuchen, besichtigen*
visitor *Besucher/in, Gast*
wait *warten (auf)*
walk *Spaziergang; (zu Fuß) gehen*
wall *Wand, Mauer*
want *wollen*
warm *warm*
wash *waschen*
watch *(Armband-)Uhr*
water *Wasser*
way *Weg, Methode, Art (und Weise)*
wear *tragen, anhaben*
weather *Wetter*
Wednesday *Mittwoch*
week *Woche*
well *gesund, gut, also*
west *Westen*
what *was, welche/r/s*
when *wenn, als, wann*
where *wo(hin)*
which *welche/r/s*
while *während*
who *wer, der/die/das, welche(r/s)*
whose *wessen, dessen*
why *warum*
with *mit, bei*
within *innerhalb (von), in*
without *ohne*
woman *Frau*
wood *Holz, Wald*
word *Wort*
work *Arbeit, Werk; funktionieren, arbeiten*
worker *Arbeiter/in*
world *Welt*
worse *schlechter, schlimmer*
worst *schlechteste, schlimmste*
write *schreiben*
wrong *falsch*
year *Jahr*
yesterday *gestern*
yet *noch*
you know *wissen Sie*
you see *sehen Sie*
young *jung*

Die neuen Wörter sind in der Reihenfolge ihres Vorkommens im Text verzeichnet.
Nicht aufgeführt sind die ca. 500 vorausgesetzten Wörter.
Die Zahl am linken Rand gibt die Seitenzahl an.

REFRESHER COURSE

Unit 1

6 **contact** ['kɒntækt] *Kontakt, Verbindung*
refresher [rɪ'freʃə] *Auffrischung*
get to know [get tə 'nəʊ] *kennen lernen*
each other [i:tʃ 'ʌðə] *einander, sich*
speech bubble ['spi:tʃ bʌbl] *Sprechblase*
material [mə'tɪəriəl] *Angaben*
box [bɒks] *Kasten, Kästchen*
group [gru:p] *Gruppe*
course [kɔ:s] *Kurs, Lehrgang*
Turkish ['tɜ:kɪʃ] *Türke/Türkin; türkisch*
Cologne [kə'ləʊn] *Köln*
electrician [ɪˌlek'trɪʃn] *Elektriker/in, Elektrotechniker/in*
plant [plɑ:nt] *Werk, Betrieb*
maintenance ['meɪntənəns] *Wartung*
technician [tek'nɪʃn] *Techniker/in*
furniture ['fɜ:nɪtʃə] *Möbel*
factory ['fæktəri] *Fabrik*
French [frentʃ] *Franzose/Französin; französisch*
Normandy ['nɔ:məndi] *Normandie*
self-employed [ˌself ɪm'plɔɪd] *selbstständig*
heating ['hi:tɪŋ] *Heizung*
engineer [ˌendʒɪ'nɪə] *Ingenieur/in, Techniker/in*
unemployed [ˌʌnɪm'plɔɪd] *arbeitslos, ohne Beschäftigung*
at the moment [ət ðə 'məʊmənt] *im Augenblick*
retrain [ˌri:'treɪn] *umschulen*
trainee [treɪ'ni:] *Auszubildende/r, Praktikant/in*
Austrian ['ɒstriən] *Österreicher/in; österreichisch*
chemist ['kemɪst] *Chemiker/in*
Afghan ['æfgæn] *Afghane/Afghanin; afghanisch*
Afghanistan [æf'gænɪstɑ:n] *Afghanistan*
Albanian [æl'beɪniən] *Albanier/in; albanisch*
Albania [æl'beɪniə] *Albanien*
Austria ['ɒstriə] *Österreich*
British ['brɪtɪʃ] *Brite/Britin; britisch*
Britain ['brɪtn] *Großbritannien*
Croatian [krəʊ'eɪʃən] *Kroate/Kroatin; kroatisch*
Croatia [krəʊ'eɪʃə] *Kroatien*
Czech [tʃek] *Tscheche/Tschechin; tschechisch*
Czech Republic [tʃek rɪ'pʌblɪk] *Tschechische Republik, Tschechien*
Dutch [dʌtʃ] *Holländer/in; holländisch*
Netherlands ['neðələndz] *Niederlande*
German ['dʒɜ:mən] *Deutsche/r; deutsch*
Germany ['dʒɜ:məni] *Deutschland*
Greek [gri:k] *Grieche/Griechin; griechisch*
Greece [gri:s] *Griechenland*
Hungarian [hʌŋ'geəriən] *Ungar/in; ungarisch*
Hungary ['hʌŋgəri] *Ungarn*
Italian [ɪ'tæljən] *Italiener/in; italienisch*
Italy ['ɪtəli] *Italien*
Kazach [kə'zæk] *Kasache/Kasachin; kasachisch*
Kazachstan [ˌkæzæk'stɑ:n] *Kasachstan*
Moroccan [mə'rɒkən] *Marokkaner/in; marokkanisch*
Morocco [mə'rɒkəʊ] *Marokko*
Polish ['pəʊlɪʃ] *Pole/Polin; polnisch*
Poland ['pəʊlənd] *Polen*
Portuguese [ˌpɔ:tʃu'gi:z] *Portugiese/Portugiesin; portugiesisch*
Portugal ['pɔ:tʃʊgl] *Portugal*
Russian ['rʌʃn] *Russe/Russin; russisch*
Russia ['rʌʃə] *Russland*
Serbian ['sɜ:biən] *Serbe/Serbin; serbisch*
Serbia ['sɜ:biə] *Serbien*
Slovak ['sləʊvæk] *Slowake/Slowakin; slowakisch*
Slovakia [sləʊ'vækiə] *Slowakei*
Spanish ['spænɪʃ] *Spanier/in; spanisch*
Spain [speɪn] *Spanien*
Ukrainian [ju'kreɪniən] *Ukrainer/in; ukrainisch*
Ukraine [ju'kreɪn] *Ukraine*
Turkey ['tɜ:ki] *Türkei*
draughtsperson ['drɑ:ftspɜ:sn] *Zeichner/in*
apprentice [ə'prentɪs] *Lehrling, Auszubildende/r*
electronics [ˌɪlek'trɒnɪks] *Elektronik*
lab assistant [ˌlæb ə'sɪstənt] *Laborassistent/in*
lathe operator ['leɪð ɒpəreɪtə] *Dreher/in*
machine operator [mə'ʃi:n ɒpəreɪtə] *Maschinenarbeiter/in*
industrial [ɪn'dʌstriəl] *Industrie-, industriell*
construction [kən'strʌkʃn] *Bau*
mechanic [mɪ'kænɪk] *Mechaniker/in*
7 **of course** [əf 'kɔ:s] *natürlich, selbstverständlich*
India ['ɪndiə] *Indien*
surname ['sɜ:neɪm] *Nach-, Familienname*
how about [ˌhaʊ ə'baʊt] *was ist mit*
echo ['ekəʊ] *Echo*
uniform ['ju:nɪfɔ:m] *Uniform*
X-ray ['eksreɪ] *Röntgenstrahl*
link [lɪŋk] *verbinden*
smoke [sməʊk] *rauchen*
go ahead [ˌgəʊ ə'hed] *fortfahren*
8 **dialogue** ['daɪəlɒg] *Dialog*
guide [gaɪd] *Anleitung*
one another [wʌn ə'nʌðə] *einander*
local ['ləʊkl] *örtlich, lokal*
enjoy [ɪn'dʒɔɪ] *gern haben/tun*
9 **identification** [aɪˌdentɪfɪ'keɪʃn] *Legitimation, Identität(sfeststellung)*
registration [ˌredʒɪ'streɪʃn] *Zulassung, Registrierung*
term [tɜ:m] *Bezeichnung, Ausdruck*
item ['aɪtəm] *Punkt, Gegenstand, Eintrag*
account [ə'kaʊnt] *Konto*
bar code ['bɑ: kəʊd] *Strichkodierung*
invoice ['ɪnvɔɪs] *Rechnung*
lottery ['lɒtəri] *Lotterie, Lotto*
postal code ['pəʊstl kəʊd] *Postleitzahl*
dictate [dɪk'teɪt] *diktieren*
infobox ['ɪnfəʊbɒks] *Infokästchen*
10 **express** [ɪk'spres] *ausdrücken*
mathematical [ˌmæθə'mætɪkl] *mathematisch*
score [skɔ:] *Ergebnis, Punkt(estand)*
game [geɪm] *Spiel*

temperature ['temprətʃə] *(erhöhte) Temperatur, Fieber*
series ['sɪəri:z] *Reihe*
common ['kɒmən] *normal, üblich*
automotive [ˌɔ:tə'məʊtɪv] *Auto(mobil)-*
limited ['lɪmɪtɪd] *mit beschränkter Haftung*
answerphone ['ɑ:nsəfəʊn] *Anrufbeantworter*
message ['mesɪdʒ] *Nachricht*
11 **calculation** [ˌkælkju'leɪʃn] *Rechenart, -operation*
basic ['beɪsɪk] *grundlegend, elementar*
work out [ˌwɜ:k 'aʊt] *ausrechnen, kalkulieren*
role [rəʊl] *Rolle*
12 **breakdown** ['breɪkdaʊn] *Panne*
broken ['brəʊkən] *kaputt, defekt*
replace [rɪ'pleɪs] *ersetzen*
highlighted ['haɪlaɪtɪd] *hervorgehoben*
correct [kə'rekt] *richtig*
hammer ['hæmə] *Hammer*
workbench ['wɜ:kbentʃ] *Werkbank*
fill in [ˌfɪl 'ɪn] *eintragen, -setzen*
13 **translate** [træns'leɪt] *übersetzen, übertragen*
14 **firm** [fɜ:m] *Firma*
Vienna [vi'enə] *Wien*
15 **order** ['ɔ:də] *Reihenfolge*
tool [tu:l] *Werkzeug*

Unit 2

16 **socket wrench** ['sɒkɪt rentʃ] *Steckschlüssel*
a pair of [ə 'peər əv] *ein Paar, eine (Zange)*
pliers ['plaɪəz] *(Kombi-)Zange*
pincers ['pɪnsəz] *Kneifzange*
Allen key ['ælən ki:] *Inbusschlüssel, Innensechskantschlüssel*
rule [ru:l] *Maß(stab), Lineal*
digital measuring slide [ˌdɪdʒɪtl 'meʒərɪŋ slaɪd] *digitaler Messschieber*
wire ['waɪə] *Draht*
round [raʊnd] *rund*
file [faɪl] *Feile*
triangular [traɪ'æŋgjələ] *dreieckig*
adjustable [ə'dʒʌstəbl] *verstellbar*
spanner ['spænə] *Schraubenschlüssel*
centre punch ['sentə pʌntʃ] *Körner*
drill [drɪl] *Bohrer, Bohrmaschine*
hacksaw ['hæksɔ:] *Bügelsäge*
metal cutters ['metl kʌtəz] *Seitenschneider*
chisel ['tʃɪzl] *Meißel, Stemmeisen*
spirit level ['spɪrɪt levl] *Wasserwaage*
jig saw ['dʒɪg sɔ:] *Stichsäge*
battery ['bætəri] *Batterie*
screwdriver ['skru:draɪvə] *Schraubenzieher, -dreher*
belt sander ['belt sænder] *Schwingschleifer*
angle grinder ['æŋgl graɪndə] *Winkelschleifer*
screw [skru:] *Schraube*
nail [neɪl] *Nagel*
17 **look at** ['lʊk ət] *ansehen, betrachten*
complete [kəm'pli:t] *vervollständigen*
wire strippers ['waɪə strɪpəz] *Abisolierzange*
column drill ['kɒləm drɪl] *Ständerbohrmaschine*
vice [vaɪs] *Schraubstock*
18 **bus stop** ['bʌs stɒp] *Bushaltestelle*
post office ['pəʊst ɒfɪs] *Postamt*
living room ['lɪvɪŋ ru:m] *Wohnzimmer*
sketch [sketʃ] *Skizze*
missing ['mɪsɪŋ] *fehlend*
shelf [ʃelf] *Regal(brett)*
drawer [drɔ:] *Schub(lade)*
soldering iron ['səʊldərɪŋ aɪən] *Lötkolben*
19 **function** ['fʌŋkʃn] *Funktion*
tin [tɪn] *Blech*
sheet [ʃi:t] *(Metall-)Platte*
drive in [ˌdraɪv 'ɪn] *einschrauben, eindrehen*
join [dʒɔɪn] *verbinden*
knock in [ˌnɒk 'ɪn] *(hin)einschlagen*
measure ['meʒə] *messen*
length [leŋθ] *Länge*
distance ['dɪstəns] *Abstand*
exact [ɪg'zækt] *genau*
thickness ['θɪknəs] *Dicke*
remove [rɪ'mu:v] *entfernen*
insulation [ˌɪnsju'leɪʃn] *Isolierung*
shape [ʃeɪp] *formen*
smooth [smu:ð] *glätten*
hole [həʊl] *Loch*
surface ['sɜ:fɪs] *(Ober-)Fläche*
pipe [paɪp] *Rohr*
20 **blunt** [blʌnt] *stumpf*
prompt [prɒmpt] *Stichwort, Vorgabe*
heavy ['hevi] *schwer*
van [væn] *Transporter*
inaccurate [ɪn'ækjərət] *ungenau*
paint brush ['peɪnt brʌʃ] *Pinsel*
knife [naɪf] *Messer*
aircraft ['eəkrɑ:ft] *Flugzeug*
21 **have to do with** [hæv tə 'du: wɪð] *zu tun haben mit*
car body ['kɑ: bɒdi] *Karosserie*
car crash ['kɑ: kræʃ] *Autounfall*
clutch [klʌtʃ] *Kupplung*
cylinder ['sɪlɪndə] *Zylinder*
fuel mix ['fju:əl mɪks] *Kraftstoffgemisch*
switch [swɪtʃ] *Schalter*
compound ['kɒmpaʊnd] *Zusammensetzung*
alcohol ['ælkəhɒl] *Alkohol*
canteen [kæn'ti:n] *Kantine, Cafeteria*
alcoholic [ˌælkə'hɒlɪk] *alkoholisch*

Unit 3

22 **ferry** ['feri] *Fähre*
Scotland ['skɒtlənd] *Schottland*
engineering [ˌendʒɪ'nɪərɪŋ] *Technik, Maschinenbau*
pier [pɪə] *Pier, Anlegestelle*
tunnel ['tʌnl] *Tunnel*
describe [dɪ'skraɪb] *beschreiben*
location [ləʊ'keɪʃn] *Lage*
between [bɪ'twi:n] *zwischen*
bank [bæŋk] *Ufer*
23 **company** ['kʌmpəni] *Unternehmen, Firma*
all over the world [ɔ:l ˌəʊvə ðə 'wɜ:ld] *in der ganzen Welt*
customer ['kʌstəmə] *Kunde/Kundin*
France [frɑ:ns] *Frankreich*
docks [dɒks] *Hafen(anlagen)*
shift [ʃɪft] *Schicht*
shift allowance [ʃɪft ə'laʊəns] *Schichtzuschlag*
working day [ˌwɜ:kɪŋ 'deɪ] *Arbeitstag*
look like ['lʊk laɪk] *aussehen wie*
typical ['tɪpɪkl] *typisch*

get up [ˌget ˈʌp] *aufstehen*
turn on [ˌtɜːn ˈɒn] *einschalten*
coffee maker [ˈkɒfi meɪkə] *Kaffeemaschine*
shower [ˈʃaʊə] *duschen*
get dressed [get ˈdrest] *sich anziehen*
journey [ˈdʒɜːni] *Fahrt*
clock on [ˌklɒk ˈɒn] *einstempeln*
day after day [ˌdeɪ ɑːftə ˈdeɪ] *tagaus, tagein*
in fact [ɪn ˈfækt] *tatsächlich*
routine [ruːˈtiːn] *Routine(arbeit)*
repair [rɪˈpeə] *reparieren*
machine [məˈʃiːn] *Maschine*
install [ɪnˈstɔːl] *installieren, einrichten*
equipment [ɪˈkwɪpmənt] *Ausrüstung, Geräte*
find out [ˌfaɪnd ˈaʊt] *feststellen*
request [rɪˈkwest] *Anfrage, Anforderung*
repair work [rɪˈpeə wɜːk] *Reparaturarbeit*
store [stɔː] *(Material-)Lager*
intranet [ˈɪntrənet] *Firmennetz(werk)*
hard copy [ˌhɑːd ˈkɒpi] *Ausdruck*
24 statement [ˈsteɪtmənt] *Aussage*
not necessarily [nɒt ˌnesəˈserəli] *nicht unbedingt, nicht notwendigerweise*
equivalent [ɪˈkwɪvələnt] *Entsprechung*
expression [ɪkˈspreʃn] *Ausdruck*
suitable [ˈsuːtəbl] *passend*
service [ˈsɜːvɪs] *durchsehen, warten*
25 sender [ˈsendə] *Absender*
check [tʃek] *(über)prüfen, kontrollieren*
per [pə] *pro, per*
mechanical [mɪˈkænɪkl] *mechanisch, technisch*
electrical [ɪˈlektrɪkl] *elektrisch, Strom-*
plumbing [ˈplʌmɪŋ] *Klempnerarbeiten*
lighting [ˈlaɪtɪŋ] *Beleuchtung*
description [dɪˈskrɪpʃn] *Beschreibung*
max. [mæks] *maximal*
character [ˈkærəktə] *Buchstabe*
priority [praɪˈɒrəti] *Vorrang, Priorität*
critical [ˈkrɪtɪkl] *kritisch*
urgent [ˈɜːdʒənt] *dringend, eilig*
automatic [ˌɔːtəˈmætɪk] *automatisch*
logistics [ləˈdʒɪstɪks] *Logistik, Auslieferung*
motor [ˈməʊtə] *Motor*
cut out [ˌkʌt ˈaʊt] *(sich) ausschalten*
meaning [ˈmiːnɪŋ] *Bedeutung*
think about [ˈθɪŋk əbaʊt] *nachdenken über*
sound [saʊnd] *Klang*
depend on [dɪˈpend ɒn] *abhängen von*
complicated [ˈkɒmplɪkeɪtɪd] *kompliziert*
spark [spɑːk] *Funke*
overheat [ˌəʊvəˈhiːt] *(sich) überhitzen, heiß laufen*
casing [ˈkeɪsɪŋ] *Gehäuse*
damage [ˈdæmɪdʒ] *beschädigen*
buckled [ˈbʌkld] *verbogen*
faulty [ˈfɔːlti] *fehlerhaft, defekt*
relay [ˈriːleɪ] *Relais*
26 rarely [ˈreəli] *selten*
seldom [ˈseldəm] *selten*
as a rule [əz ə ˈruːl] *in der Regel*
generally [ˈdʒenrəli] *im Allgemeinen*
now and then [ˌnaʊ ənd ˈðen] *ab und zu, hin und wieder*
regularly [ˈregjələli] *regelmäßig*
housework [ˈhaʊswɜːk] *Hausarbeit(en)*
qualified [ˈkwɒlɪfaɪd] *qualifiziert, ausgebildet*
27 parents [ˈpeərənts] *Eltern*
made in [ˈmeɪd ɪn] *hergestellt in*
copier [ˈkɒpiə] *Kopierer*
cycling [ˈsaɪklɪŋ] *Radfahren*
girlfriend [ˈgɜːlfrend] *Freundin*
careful [ˈkeəfl] *sorgfältig*
overtime [ˈəʊvətaɪm] *Überstunden*

Unit 4

28 accident [ˈæksɪdənt] *Unfall*
safety [ˈseɪfti] *Sicherheit, Arbeitsschutz*
officer [ˈɒfɪsə] *Beauftragte/r, Chef/in*
installation [ˌɪnstəˈleɪʃn] *Installation*
hurt [hɜːt] *verletzt*
scaffold tower [ˈskæfəʊld taʊə] *(Bau-)Gerüst*
cause [kɔːz] *verursachen*
mistake [mɪˈsteɪk] *Fehler*
at fault [ət ˈfɔːlt] *schuldig*
mobile [ˈməʊbaɪl] *mobil, beweglich*
guard rail [ˈgɑːd reɪl] *Schutzgeländer*
safety harness [ˈseɪfti hɑːnɪs] *Sicherheitsgurt*
warning sign [ˈwɔːnɪŋ saɪn] *Warnzeichen, -schild*
hard hat [ˌhɑːd ˈhæt] *Schutzhelm*
stabilizer [ˈsteɪbəlaɪzə] *Stabilisator*
cone [kəʊn] *Warnhütchen*
locking castor [ˌlɒkɪŋ ˈkɑːstə] *feststellbare Laufrolle*
surround [səˈraʊnd] *umgeben*
support [səˈpɔːt] *ab-, unterstützen*
fix [fɪks] *befestigen*
corner [ˈkɔːnə] *Ecke*
mark [mɑːk] *markieren, kennzeichnen*
catch [kætʃ] *(auf)fangen*
height [haɪt] *Höhe*
sway [sweɪ] *schwingen, schwanken*
close [kləʊs] *nahe*
29 floor [flɔː] *(Fuß-)Boden*
forklift (truck) [ˈfɔːklɪft] *Gabelstapler*
hit [hɪt] *gegen etw fahren, zusammenstoßen mit etw*
pre-assembly [ˌpriː əˈsembli] *Vormontage*
assembly [əˈsembli] *Montage, Zusammenbau*
p.m. [ˌpiː ˈem] *nachmittags*
go on [ˌgəʊ ˈɒn] *fortfahren*
swing door [ˌswɪŋ ˈdɔː] *Schwingtür*
happen [ˈhæpən] *passieren, geschehen*
injury [ˈɪndʒəri] *Verletzung*
guy [gaɪ] *Typ, Junge*
insured [ɪnˈʃʊəd] *versichert*
guarantee [ˌgærənˈtiː] *garantieren*
environment [ɪnˈvaɪrənmənt] *Umgebung, Umfeld*
measure [ˈmeʒə] *Maßnahme*
promise [ˈprɒmɪs] *versprechen*
as well [əz ˈwel] *auch, ebenfalls*
miss [mɪs] *übersehen*
concrete [ˈkɒŋkriːt] *Beton*
what else [wɒt ˈels] *was sonst (noch)*
for God's sake [fə gɒdz ˈseɪk] *um Gottes willen*
trouble [ˈtrʌbl] *Ärger, Mühe, Problem/e*
30 in this case [ɪn ðɪs ˈkeɪs] *in diesem Fall*
lock [lɒk] *feststellen, arretieren*
suddenly [ˈsʌdnli] *plötzlich*
move [muːv] *(sich) bewegen*
bad luck [bæd ˈlʌk] *Unglück, Pech*

in front of [ɪn 'frʌnt əv] *vor*
swerve [swɜ:v] *einen Schlenker machen, ausschwenken*
kiss [kɪs] *leicht berühren*
report [rɪ'pɔ:t] *Bericht*
insurer [ɪn'ʃʊərə] *Versicherer, Versicherung*
get on with [ˌget 'ɒn wɪð] *erledigen*
at once [ət 'wʌns] *sofort*
quality ['kwɒləti] *Qualität*
31 **factual** ['fæktʃuəl] *sachlich*
extract ['ekstrækt] *Auszug*
involve [ɪn'vɒlv] *beteiligen*
inspector [ɪn'spektə] *Kontrolleur/in*
safety measure ['seɪfti meʒə] *Sicherheitsmaßnahme*
site [saɪt] *Baustelle*
flash [flæʃ] *blinken*
yellow ['jeləʊ] *gelb*
safety net ['seɪfti net] *Sicherheitsnetz*
locking jack ['lɒkɪŋ dʒæk] *Karabinerhaken*
platform ladder ['plætfɔ:m lædə] *Trittleiter*
locate [ləʊ'keɪt] *platzieren*
approx (approximately) [ə'prɒksɪmətli] *ungefähr, zirka*
avoid [ə'vɔɪd] *ausweichen*
grid [grɪd] *Tabelle*
figure ['fɪgə] *Zahl, Ziffer*
step [step] *Stufe*
32 **day before yesterday** [deɪ bɪˌfɔ: 'jestədeɪ] *vorgestern*
Easter ['i:stə] *Ostern*
in those days [ɪn ðəʊz 'deɪz] *damals, in jenen Tagen*
33 **worksheet** ['wɜ:kʃi:t] *Arbeitsbogen*
swap [swɒp] *tauschen*
at least [ət 'li:st] *mindestens*
department [dɪ'pɑ:tmənt] *Abteilung*
timesheet ['taɪmʃi:t] *(Zeit-)Kontrollliste*
window ['wɪndəʊ] *Fenster*
light bulb ['laɪt bʌlb] *Glühbirne*
clean [kli:n] *reinigen, säubern*
filter ['fɪltə] *Filter*
extractor fan [ɪk'stræktə fæn] *Entlüfter*
conveyor [kən'veɪə] *Aufzug*
fetch [fetʃ] *holen*
spare part [ˌspeə 'pɑ:t] *Ersatzteil*
unfortunately [ʌn'fɔ:tʃənətli] *unglücklicherweise, leider*
save [seɪv] *sparen*
climb [klaɪm] *klettern, steigen*
all the time [ˌɔ:l ðə 'taɪm] *die ganze Zeit*
obey [ə'beɪ] *befolgen*
just for fun [dʒʌst fə 'fʌn] *nur zum Spaß*
be in charge of [ɪn 'tʃɑ:dʒ əv] *die Verantwortung tragen für*

Unit 5

34 **street plan** ['stri:t plæn] *Straßenkarte*
bottom ['bɒtəm] *(unteres) Ende, unten*
delivery [dɪ'lɪvəri] *Zustellung, Lieferung*
satnav ['sætnæv] *Navi(gationsgerät)*
hopeless ['həʊpləs] *hoffnungslos*
strange [streɪndʒ] *fremd*
luckily ['lʌkɪli] *glücklicherweise*
directions [də'rekʃnz] *Wegbeschreibung*
reach [ri:tʃ] *erreichen*
football ground ['fʊtbɔ:l graʊnd] *Fußballplatz*
bus station ['bʌs steɪʃn] *Busbahnhof*
pedestrian crossing [pəˌdestriən 'krɒsɪŋ] *Fußgängerüberweg*
35 **instruction** [ɪn'strʌkʃn] *Anweisung*
straight on [ˌstreɪt 'ɒn] *geradeaus*
red [red] *rot*
gate [geɪt] *Tor*
turn [tɜ:n] *abbiegen*
roundabout ['raʊndəbaʊt] *Kreisverkehr, Kreisel*
immediate [ɪ'mi:diət] *sofort, unmittelbar*
no entry [nəʊ 'entri] *Durchfahrt verboten*
T-junction ['ti: dʒʌŋkʃn] *T-Kreuzung*
one-way system [ˌwʌn weɪ 'sɪstəm] *Einbahnsystem, -verkehr*
traffic ['træfɪk] *(Straßen-)Verkehr*
traffic lights ['træfɪk laɪts] *(Verkehrs-)Ampel*
repeat [rɪ'pi:t] *wiederholen*
36 **awful** ['ɔ:fl] *furchtbar*
along [ə'lɒŋ] *entlang*
don't mention it ['menʃn] *nichts zu danken*
you're welcome [jɔ: 'welkəm] *gern geschehen*
green [gri:n] *grün*
good luck [ˌgʊd 'lʌk] *viel Glück*
ask the way [ɑ:sk ðə 'weɪ] *nach dem Weg fragen*
town hall [ˌtaʊn 'hɔ:l] *Rathaus*
technical college ['teknɪkl kɒlɪdʒ] *technische Hoch-, Fachschule*
church [tʃɜ:tʃ] *Kirche*
police station [pə'li:s steɪʃn] *Polizeiwache, -revier*
youth hostel ['ju:θ hɒstl] *Jugendherberge*
on foot [ɒn 'fʊt] *zu Fuß*
lane [leɪn] *Gasse, Weg*
37 **shopping mall** ['ʃɒpɪŋ mɔ:l] *Einkaufszentrum, -passage*
car park ['kɑ: pɑ:k] *Parkplatz, Parkhaus*
leisure centre ['leʒə sentə] *Freizeitzentrum*
tourist information office [ˌtʊərɪst ɪnfə'meɪʃn ɒfɪs] *Touristeninformationsbüro*
turn left/right [tɜ:n 'left, 'raɪt] *links/rechts abbiegen*
on the left/right [ɒn ðə 'left, 'raɪt] *auf der linken/rechten Seite*
on your left/right [ɒn jɔ: 'left, 'raɪt] *zu Ihrer Linken/Rechten*
various ['veəriəs] *verschieden*
multiplex ['mʌltɪpleks] *Kinozentrum*
screen [skri:n] *Leinwand*
supertram ['su:pətræm] *Großraum-Straßenbahn*
in the end [ɪn ði 'end] *zum Schluss, schließlich*
38 **leaflet** ['li:flət] *Broschüre, Prospekt*
gap [gæp] *Lücke*
boast [bəʊst] *prahlen, angeben*
close [kləʊz] *(ab)schließen*
thief [θi:f] *Dieb/in*
Dos and Don'ts [ˌdu:z ən 'dəʊnts] *Gebote und Verbote*
vehicle ['vi:əkl] *Fahrzeug*
stranger ['streɪndʒə] *Fremde/r*
camera ['kæmərə] *Fotoapparat, Kamera*
document ['dɒkjumənt] *Dokument, Papier*
sunroof ['sʌnru:f] *Schiebedach*
lonely ['ləʊnli] *einsam*
wonderful ['wʌndəfl] *wunderbar*
attendant [ə'tendənt] *Wächter/in*

39 **area** ['eəriə] *Bereich*
building ['bɪldɪŋ] *Gebäude, Bau*

Unit 6

40 **during** ['djʊərɪŋ] *während*
logical ['lɒdʒɪkl] *logisch*
in practice [ɪn 'præktɪs] *in der Praxis*
goods inward [gʊdz 'ɪnwəd] *Warenannahme*
progress ['prəʊgres] *Verfahren, Weg*
inspection [ɪn'spekʃn] *Kontrolle*
packaging ['pækɪdʒɪŋ] *Verpackung*
dispatch [dɪ'spætʃ] *Versand, Auslieferung*
process ['prəʊses] *Prozess, Ablauf*
drawing ['drɔ:ɪŋ] *Zeichnung*
administration [əd,mɪnɪ'streɪʃn] *Verwaltung*
security [sɪ'kjʊərəti] *Sicherheit*
caption ['kæpʃn] *Bildunterschrift*
object ['ɒbdʒɪkt] *Gegenstand, Objekt*
fit [fɪt] *passen*
component [kəm'pəʊnənt] *(Einzel-)Teil*
cardboard ['kɑ:dbɔ:d] *Pappe*
exhaust outlet [ɪg,zɔ:st 'aʊtlet] *Auspuff(rohr)*
shelving ['ʃelvɪŋ] *Regale*
suitcase ['su:tkeɪs] *Koffer*
surveillance camera [sɜ:'veɪləns kæmərə] *Überwachungskamera*
tap [tæp] *(Wasser-)Hahn*
41 **take place** [teɪk 'pleɪs] *stattfinden*
deliver [dɪ'lɪvə] *anliefern*
suspicious [sə'spɪʃəs] *verdächtig*
after-sales [,ɑ:ftə 'seɪlz] *Kundendienst-*
service ['sɜ:vɪs] *(Kunden-)Dienst*
data processing [,deɪtə 'prəʊsesɪŋ] *Datenverarbeitung, EDV*
loading ramp ['ləʊdɪŋ ræmp] *Laderampe*
management ['mænɪdʒmənt] *Geschäftsleitung, Verwaltung*
quality assurance ['kwɒləti əʃʊərəns] *Qualitätssicherung*
reception [rɪ'sepʃn] *EmpfangRezeption*
goods outward [gʊdz 'aʊtwəd] *Warenauslieferung*
workshop ['wɜ:kʃɒp] *Werkstatt, Werkshalle*
research and development [rɪ,sɜ:tʃ ən dɪ'veləpmənt] *Forschung und Entwicklung*
42 **training** ['treɪnɪŋ] *Ausbildung, Schulung*
apprenticeship [ə'prentɪʃɪp] *Lehre, Ausbildung(splatz)*
decide [dɪ'saɪd] *entscheiden*
disappoint [,dɪsə'pɔɪnt] *enttäuschen*
interest ['ɪntrəst] *Interesse*
applicant ['æplɪkənt] *Bewerber/in*
in the middle of [ɪn ðə 'mɪdl əv] *mitten in/auf*
yard [jɑ:d] *Hof*
truck [trʌk] *Last(kraft)wagen, Laster, Lastzug*
make up [,meɪk 'ʌp] *bilden*
assembly plant [ə'sembli plɑ:nt] *Montage-, Herstellwerk*
turn round [,tɜ:n 'raʊnd] *(sich) umdrehen*
anyhow ['enihaʊ] *jedenfalls*
-shaped [ʃeɪpt] *-förmig*
made up of [,meɪd 'ʌp əv] *bestehend aus*
produce [prə'dju:s] *produzieren, herstellen*
improve [ɪm'pru:v] *verbessern*
production planning [prə'dʌkʃn plænɪŋ] *Produktionsplanung*
except [ɪk'sept] *außer*
any more [,eni 'mɔ:] *noch irgendwelche*
so far [,səʊ 'fɑ:] *bis jetzt*
across [ə'krɒs] *(quer) über*
keep to ['ki:p tə] *sich halten an*
walkway ['wɔ:kweɪ] *Gang, Fußweg*
all right [ɔ:l 'raɪt] *verstanden*
supplier [sə'plaɪə] *Lieferant/in, Zulieferer*
thorough ['θʌrə] *gründlich*
inspect [ɪn'spekt] *inspizieren, überprüfen*
sort [sɔ:t] *sortieren*
store [stɔ:] *lagern*
wonder ['wʌndə] *sich fragen*
manufacturing [,mænju'fæktʃərɪŋ] *Produktion, Fertigung*
at all [ət 'ɔ:l] *überhaupt*
tough [tʌf] *hart, schwierig*
competitive [kəm'petətɪv] *konkurrenzfähig*
global ['gləʊbl] *global, weltweit*
marketplace ['mɑ:kɪtpleɪs] *Markt(platz)*
43 **collect** [kə'lekt] *(ab)holen*
double doors [,dʌbl 'dɔ:z] *Doppeltür*
video monitor [,vɪdiəʊ 'mɒnɪtə] *Bildschirm*
if you like [ɪf ju 'laɪk] *wenn Sie wollen/mögen*
boss [bɒs] *Chef/in*
angry ['æŋgri] *böse, wütend, zornig*
skilled [skɪld] *qualifiziert*
cook [kʊk] *kochen*
serve [sɜ:v] *servieren*
washing-up [,wɒʃɪŋ 'ʌp] *Abwasch, Spülen*
computer-aided design [kəm,pju:tər eɪdɪd dɪ'zaɪn] *computergestützte Konstruktion*
suppose [sə'pəʊz] *glauben, annehmen*
efficient [ɪ'fɪʃnt] *leistungsfähig*
surprised [sə'praɪzd] *überrascht*
search [sɜ:tʃ] *durchsuchen*
threatening ['θretnɪŋ] *bedrohlich*
get results [get rɪ'zʌlts] *Erfolg haben*
willing ['wɪlɪŋ] *bereit(willig)*
reject [rɪ'dʒekt] *ablehnen*
properly ['prɒpəli] *richtig*
44 **smell** [smel] *duften, riechen*
taste [teɪst] *schmecken*
unpack [,ʌn'pæk] *auspacken*
recorder [rɪ'kɔ:də] *Rekorder*
contents ['kɒntents] *Inhalt*
carton ['kɑ:tn] *Karton*
packing ['pækɪŋ] *Pack-*
incomplete [,ɪnkəm'pli:t] *unvollständig*
dealer ['di:lə] *Händler/in*
direct [də'rekt] *direkt*
sunlight ['sʌnlaɪt] *Sonnenlicht*
connect [kə'nekt] *anschließen*
SCART jack ['skɑ:t dʒæk] *SCART-Stecker*
cable ['keɪbl] *Kabel*
force [fɔ:s] *Gewalt*
power cable ['paʊə keɪbl] *Stromkabel*
power point ['paʊə pɔɪnt] *Elektroanschluss, Steckdose*
protect [prə'tekt] *schützen*
button ['bʌtn] *Taste, Knopf*
remote control [rɪ,məʊt kən'trəʊl] *Fernbedienung*
handset ['hændset] *Handapparat*
45 **instruct** [ɪn'strʌkt] *ein-, anweisen*
football match ['fʊtbɔ:l mætʃ] *Fußballspiel*
switch off [,swɪtʃ 'ɒf] *ausschalten*

MAIN COURSE

Unit 1

46 **abroad** [ə'brɔ:d] *im Ausland*
experience [ɪk'spɪəriəns] *(Berufs-)Erfahrung*
career chances [kə'rɪə tʃɑ:nsɪz] *Berufschancen*
global village [ˌgləʊbl 'vɪlɪdʒ] *globales Dorf*
ad [æd] *Anzeige*
Irish ['aɪrɪʃ] *irisch*
headquarters [ˌhed'kwɔ:təz] *Firmensitz, Zentrale*
realia [reɪ'ɑ:liə] *Realien*
job centre ['dʒɒb sentə] *Arbeitsagentur*
job offer ['dʒɒb ɒfə] *Stellenangebot*
careers advice [kəˌrɪəz əd'vaɪs] *Berufsberatung*
employment agency [ɪm'plɔɪmənt eɪdʒənsi] *(private) Stellenvermittlung*
47 **Ireland** ['aɪələnd] *Irland*
successful [sək'sesfl] *erfolgreich*
manufacturer [ˌmænju'fæktʃərə] *Hersteller/in, Produzent/in*
advanced [əd'vɑ:nst] *fortgeschritten*
device [dɪ'vaɪs] *Vorrichtung, Gerät*
owing to ['əʊɪŋ tə] *wegen, aufgrund*
growth [grəʊθ] *Wachstum, Zunahme*
export ['ekspɔ:t] *Ausfuhr, Export*
throughout [θru:'aʊt] *überall (in), in ganz*
market leader [ˌmɑ:kɪt 'li:də] *Marktführer*
rapid ['ræpɪd] *schnell, rasch*
expand [ɪk'spænd] *sich ausdehnen, expandieren*
European [ˌjʊərə'pi:ən] *europäisch*
employ [ɪm'plɔɪ] *beschäftigen*
head office [ˌhed 'ɒfɪs] *Zentrale*
associate [ə'səʊʃiət] *Partner/in*
offer ['ɒfə] *bieten*
reward [rɪ'wɔ:d] *Lohn, Vergütung*
generous ['dʒenərəs] *großzügig*
salary ['sæləri] *Gehalt, Lohn*
annual ['ænjuəl] *jährlich*
performance bonus [pə'fɔ:məns bəʊnəs] *Leistungsprämie*
vacation [və'keɪʃn] *Urlaub, Ferien*
health insurance ['helθ ɪnʃʊərəns] *Krankenversicherung*
insurance [ɪn'ʃʊərəns] *Versicherung*
pension plan ['penʃn plæn] *betriebliche Altersversorgung, Betriebsrente*
attraction [ə'trækʃn] *Anreiz*
receive [rɪ'si:v] *erhalten*
home country [ˌhəʊm 'kʌntri] *Heimatland*
major ['meɪdʒə] *bedeutend, groß*
initiative [ɪ'nɪʃətɪv] *Initiative*
in brief [ɪn 'bri:f] *kurz (zusammengefasst)*
payment ['peɪmənt] *Bezahlung*
perk [pɜ:k] *zusätzliche Leistung*
model ['mɒdl] *Modell*
category ['kætəgəri] *Kategorie, Fachgebiet*
post [pəʊst] *aufgeben*
Human Resources (HR) [ˌhju:mən rɪ'sɔ:sɪz] *Personalabteilung*
48 **do well** [ˌdu: 'wel] *gut abschneiden, gut gehen*
terrorism ['terərɪzəm] *Terrorismus*
crime [kraɪm] *Kriminalität*
motion detector ['məʊʃn dɪtektə] *Bewegungsmelder*
change [tʃeɪndʒ] *(sich) verändern*
industry ['ɪndəstri] *Industrie*
in brackets [ɪn 'brækɪts] *in Klammern*
roughly ['rʌfli] *ungefähr*
application [ˌæplɪ'keɪʃn] *Bewerbung*
apply [ə'plaɪ] *bewerben*
single ['sɪŋgl] *alleinstehend, unverheiratet*
contribution [ˌkɒntrɪ'bju:ʃn] *Beitrag*
at present [ət 'preznt] *momentan, gegenwärtig*
fully ['fʊli] *völlig*
advertisement [əd'vɜ:tɪsmənt] *Anzeige*
attach [ə'tætʃ] *beilegen, -fügen*
hope for ['həʊp fə] *hoffen auf*
favourable ['feɪvərəbl] *positiv, günstig*
Yours sincerely [jɔ:z sɪn'sɪəli] *Mit freundlichen Grüßen*
change [tʃeɪndʒ] *(Ver-)Änderung*
personal ['pɜ:sənl] *persönlich*
expectation [ˌekspek'teɪʃn] *Erwartung*
49 **CV** [ˌsi: 'vi:] *Lebenslauf*
address [ə'dres] *Adresse*
birth [bɜ:θ] *Geburt*
daytime ['deɪtaɪm] *tagsüber*
phone number ['fəʊn nʌmbə] *Telefonnummer*
education [ˌedʒu'keɪʃn] *(Schul-)Bildung, Ausbildung*
certificate [sə'tɪfɪkət] *Zeugnis*
university [ˌju:nɪ'vɜ:səti] *Universität, Hochschule*
applied science [əˌplaɪd 'saɪəns] *angewandte Wissenschaft*
average ['ævərɪdʒ] *durchschnittlich*
grade [greɪd] *Note*
unsatisfactory [ˌʌnˌsætɪs'fæktəri] *unbefriedigend, ungenügend*
comprehensive school [ˌkɒmprɪ'hensɪv sku:l] *Gesamtschule*
junior certificate [ˌdʒu:niə sə'tɪfɪkət] *Mittlere Reife*
work experience ['wɜ:k ɪkspɪəriəns] *Berufs-, Arbeitserfahrung, Praktikum*
generation [ˌdʒenə'reɪʃn] *Erzeugung*
to date [tə 'deɪt] *bis jetzt*
skills [skɪlz] *Fertigkeiten*
member ['membə] *Mitglied, Angehörige/r*
football club ['fʊtbɔ:l klʌb] *Fußballverein*
circuit ['sɜ:kɪt] *(Strom-)Kreis*
driving licence ['draɪvɪŋ laɪsns] *Führerschein*
class [klɑ:s] *Klasse*
reference ['refrəns] *Referenz, (Arbeits-)Zeugnis, Empfehlungsschreiben*
supply [sə'plaɪ] *liefern*
upon request [əˌpɒn rɪ'kwest] *auf Anfrage*
50 **occupation** [ˌɒkju'peɪʃn] *Beruf, Tätigkeit*
clear [klɪə] *klar, deutlich*
clue [klu:] *Hinweis*
outside [ˌaut'saɪd] *außerhalb*
conference room ['kɒnfərəns ru:m] *Besprechungszimmer, Konferenzraum*
nice to meet you [ˌnaɪs tə 'mi:t ju] *schön, Sie kennen zu lernen*
take a seat [ˌteɪk ə 'si:t] *Platz nehmen*
duty ['dju:ti] *Aufgabe, Pflicht*
51 **such as** ['sʌtʃ əz] *wie (zum Beispiel)*
what about ['wɒt əbaʊt] *was ist mit?*
occasionally [ə'keɪʒənəli] *gelegentlich*
precise [prɪ'saɪs] *genau, präzise*

top of the class [ˌtɒp əv ðə 'klɑːs] *Klassenbeste/r*
overqualified [ˌəʊvə'kwɒlɪfaɪd] *überqualifiziert*
in spite of [ɪn 'spaɪt əv] *trotz*
variety [və'raɪəti] *Abwechslung*
prospect ['prɒspekt] *Aussicht, Chance*
let's see [lets 'siː] *mal sehen*
unusual [ʌn'juːʒuəl] *ungewöhnlich, unüblich*
pay one's share [peɪ wʌnz 'ʃeə] *seinen Anteil zahlen*
unless [ən'les] *wenn nicht, es sei denn*
workplace ['wɜːkpleɪs] *Arbeitsplatz*
52 **first of all** [fɜːst əv 'ɔːl] *zu(aller)erstzunächst*
machinery [mə'ʃiːnəri] *Maschinen, Geräte*
general ['dʒenrəl] *allgemein*
exam [ɪg'zæm] *Prüfung, Examen*
contractor [kən'træktə] *Handwerksbetrieb, Bauunternehmen*
choice [tʃɔɪs] *Wahl, Auswahl*
over the top [ˌəʊvə ðə 'tɒp] *zusätzlich, übertrieben*
discuss [dɪ'skʌs] *besprechen, diskutieren (über)*
agree [ə'griː] *übereinstimmen*
reaction [ri'ækʃn] *Reaktion*
too good to be true [tuː ˌgʊd tə bi 'truː] *zu gut, um wahr zu sein*
53 **refer to** [rɪ'fɜː tə] *sprechen von, hinweisen auf*
grow [grəʊ] *wachsen*
comment ['kɒment] *Bemerkung*
react [ri'ækt] *reagieren*
for example [fər ɪg'zɑːmpl] *zum Beispiel*
weakness ['wiːknəs] *Schwäche*
obvious ['ɒbviəs] *offensichtlich*
clever ['klevə] *klug*
bored [bɔːd] *gelangweilt*
demanding [dɪ'mɑːndɪŋ] *anspruchsvoll*
go out in the field [gəʊ ˌaʊt ɪn ðə 'fiːld] *ins Freie gehen*
independent [ˌɪndɪ'pendənt] *unabhängig*
in my opinion [ɪn maɪ ə'pɪnjən] *meiner Meinung nach*
54 **everyday life** [ˌevrideɪ 'laɪf] *Alltag*
argue ['ɑːgjuː] *(sich) streiten*
chill out [ˌtʃɪl 'aʊt] *sich entspannen, abhängen*
go clubbing [gəʊ 'klʌbɪŋ] *in Diskos/Nachtklubs gehen*
dancing ['dɑːnsɪŋ] *Tanzen*
movie ['muːvi] *Kino*
pub [pʌb] *Kneipe, Lokal*
hang around [ˌhæŋ ə'raʊnd] *herumhängen*
text message ['tekst mesɪdʒ] *SMS(-Nachricht)*
55 **look for** ['lʊk fə] *suchen nach*
English-speaking ['ɪŋglɪʃspiːkɪŋ] *englischsprachig*
tyre ['taɪə] *Reifen*
engine ['endʒɪn] *Motor*
oil [ɔɪl] *Öl*
fuel tank ['fjuːəl tæŋk] *Treibstofftank*
put away [ˌpʊt ə'weɪ] *weglegen*
56 **interview** ['ɪntəvjuː] *interviewen*
spend [spend] *verbringen*
57 **question tag** ['kwestʃən tæg] *Frageanhängsel*

Unit 2

58 **continent** ['kɒntɪnənt] *(europäischer) Kontinent*
belong [bɪ'lɒŋ] *gehören*
wordfield ['wɜːdfiːld] *Wortfeld*
just the thing [dʒʌst ðə 'θɪŋ] *genau das Richtige*
car ferry ['kɑː feri] *Autofähre*
comfortable ['kʌmftəbl] *bequem*
up to ['ʌp tə] *bis zu*
quarter ['kwɔːtə] *Viertel*
though [ðəʊ] *obwohl*
Rhine [raɪn] *Rhein*
59 **via** ['vaɪə] *über*
break [breɪk] *Pause*
cross [krɒs] *überqueren*
border ['bɔːdə] *Grenze*
land [lænd] *landen*
luggage ['lʌgɪdʒ] *Gepäck*
ferry port ['feri pɔːt] *Fährhafen*
eastern ['iːstən] *Ost-*
originally [ə'rɪdʒənəli] *ursprünglich*
travel plans ['trævl plænz] *Reisepläne*
expect [ɪk'spekt] *erwarten*
60 **sail** [seɪl] *ablegen, den Hafen verlassen*
calculate ['kælkjuleɪt] *berechnen*
at first [ət 'fɜːst] *zuerst, anfangs*
half past seven [ˌhɑːf pɑːst 'sevn] *halb acht*
exit ['eksɪt] *Ausfahrt, Ausgang*
motorway ['məʊtəweɪ] *Autobahn*
busy ['bɪzi] *besetzt, belebt*
cash desk ['kæʃ desk] *Kasse*
queue [kjuː] *(Warte-)Schlange*
come on [ˌkʌm 'ɒn] *beeil dich*
I don't have all day [aɪ dəʊnt həv ˌɔːl 'deɪ] *ich habe nicht den ganzen Tag Zeit*
at last [ət 'lɑːst] *schließlich*
chat [tʃæt] *sich unterhalten*
relax [rɪ'læks] *(sich) entspannen*
after all [ˌɑːftər 'ɔːl] *schließlich*
stop and go [ˌstɒp ənd 'gəʊ] *schleppender Verkehr*
access ramp ['ækses ræmp] *Zufahrtsrampe*
filter ['fɪltə] *einspurige Auffahrt, Abbiegespur*
operate ['ɒpəreɪt] *in Betrieb sein*
at a time [ət ə 'taɪm] *auf einmal*
61 **tailback** ['teɪlbæk] *Autoschlange, Rückstau*
rush-hour ['rʌʃ aʊə] *Hauptverkehrszeit, Berufsverkehr*
fool [fuːl] *Dummkopf*
dashboard ['dæʃbɔːd] *Armaturenbrett*
flick over [ˌflɪk 'əʊvə] *weiterrücken*
maximum ['mæksɪməm] *Höchst-*
speed [spiːd] *Geschwindigkeit*
result [rɪ'zʌlt] *Resultat, Ergebnis*
police car [pə'liːs kɑː] *Polizeiwagen*
ambulance ['æmbjələns] *Krankenwagen*
overtake [ˌəʊvə'teɪk] *überholen*
hard shoulder [ˌhɑːd 'ʃəʊldə] *Seitenstreifen, Standspur*
direction [də'rekʃn] *Richtung*
siren ['saɪrən] *Sirene*
scream [skriːm] *heulen*
warning light ['wɔːnɪŋ laɪt] *Warnlicht*
mobile phone [ˌməʊbaɪl 'fəʊn] *Handy*
62 **uneasy** [ʌn'iːzi] *unruhig, unbehaglich*
realise ['rɪəlaɪz] *erkennen, feststellen*
park [pɑːk] *parken*
terminal ['tɜːmɪnl] *Abfertigungsgebäude*
continue [kən'tɪnjuː] *fortsetzen*
single ['sɪŋgl] *einfach(e Fahrt)*
motorist ['məʊtərɪst] *Autofahrer/in*

fare [feə] *Fahrpreis*
price [praɪs] *Preis*
economy [ɪ'kɒnəmi] *Touristenklasse*
return [rɪ'tɜ:n] *Rück(reise)*
appropriate [ə'prəʊpriət] *entsprechend*
available [ə'veɪləbl] *verfügbar*
route [ru:t] *Route, Strecke*
craft [krɑ:ft] *Fahrzeug, Schiff*
overnight [,əʊvə'naɪt] *über Nacht, nachts*
63 **revise** [rɪ'vaɪz] *revidieren*
arrangement [ə'reɪndʒmənt] *Termin, Vereinbarung*
period ['pɪəriəd] *Periode*
allow [ə'laʊ] *rechnen mit*
act out [,ækt 'aʊt] *vorspielen, aufführen*
64 **roadworks** ['rəʊdwɜ:ks] *Straßenbauarbeiten*
warn [wɔ:n] *warnen, darauf hinweisen*
lose [lu:z] *verlieren*
speed limit ['spi:d lɪmɪt] *Geschwindigkeitsbegrenzung*
annoyed [ə'nɔɪd] *verärgert*
worry ['wʌri] *sich Sorgen machen*
diary ['daɪəri] *Terminkalender, Tagebuch*
renew [rɪ'nju:] *erneuern*
adjust [ə'dʒʌst] *einstellen*
brakes [breɪks] *Bremsen*
steering ['stɪərɪŋ] *Lenkung*
anti-freeze ['ænti fri:z] *Frostschutz(mittel)*
windscreen wiper ['wɪndskri:n waɪpə] *Scheibenwischer*
carry out [,kæri 'aʊt] *durchführen*
road test ['rəʊd test] *Straßentest*
65 **rain** [reɪn] *regnen*
police officer [pə'li:s ɒfɪsə] *Polizeibeamter/beamtin*
be unable to [bi ʌn'eɪbl tə] *nicht können*
test the nerves of [test ðə 'nɜ:vs əv] *die Nerven strapazieren*
on one's way to [ɒn wʌnz 'weɪ tə] *auf dem Weg zur*
condition [kən'dɪʃn] *Zustand*
shine [ʃaɪn] *scheinen*
dazzle ['dæzl] *blenden*
drug [drʌg] *Droge, Rauschgift*
fast lane [,fɑ:st 'leɪn] *Überholspur*
neither ['naɪðə] *keiner der*
set off [,set 'ɒf] *in Gang setzen*
chain of events [tʃeɪn əv ɪ'vents] *Kette von Ereignissen*
block [blɒk] *blockieren*
in time [ɪn 'taɪm] *rechtzeitig*
according to [ə'kɔ:dɪŋ tə] *nach Aussage von*
witness ['wɪtnəs] *Zeuge/Zeugin*
far too ['fɑ: tu:] *bei weitem*

Unit 3

66 **settle in** [,setl 'ɪn] *sich eingewöhnen, sich zurechtfinden*
noisy ['nɔɪzi] *laut*
write in full [raɪt ɪn 'fʊl] *ausschreiben*
abbreviation [ə,bri:vi'eɪʃn] *Abkürzung*
compact [kəm'pækt] *kompakt, klein*
bedsitter [,bed'sɪtə] *möbliertes Zimmer*
shared [ʃeəd] *gemeinsam*
bathroom ['bɑ:θru:m] *Bad(ezimmer)*
ideal [aɪ'di:əl] *ideal*
deposit [dɪ'pɒzɪt] *Anzahlung, Kaution*
call [kɔ:l] *anrufen*
weekday ['wi:kdeɪ] *Wochentag, Werktag*
convert [kən'vɜ:t] *umbauen*
Victorian [vɪk'tɔ:riən] *viktorianisch*
furnished ['fɜ:nɪʃt] *möbliert, eingerichtet*
garage ['gærɑ:ʒ] *Garage*
unfurnished [ʌn'fɜ:nɪʃt] *unmöbliert*
cottage ['kɒtɪdʒ] *kleines (Land-)Haus, Ferienhaus*
resident ['rezɪdənt] *Bewohner/in*
couple ['kʌpl] *Ehepaar*
renovate ['renəveɪt] *renovieren*
houseboat ['haʊsbəʊt] *Hausboot*
living area [,lɪvɪŋ 'eəriə] *Wohnfläche*
mains services ['meɪnz sɜ:vɪsɪz] *Versorgungsanschlüsse*
tax [tæks] *Steuer*
67 **think up** [,θɪŋk 'ʌp] *(sich) ausdenken*
similar ['sɪmələ] *ähnlich*
illustrate ['ɪləstreɪt] *illustrieren, zeigen*
accommodation [ə,kɒmə'deɪʃn] *Unterkunft*
definitely ['defɪnətli] *bestimmt, sicherlich*
modern ['mɒdn] *modern*
all modern conveniences [ɔ:l ,mɒdn kən'vi:niənsɪz] *aller Komfort*
68 **entrance** ['entrəns] *Eingang*
terrible ['terəbl] *schrecklich, furchtbar*
appointment [ə'pɔɪntmənt] *Termin, Verabredung*
welcome to ['welkəm tə] *willkommen bei*
name tag ['neɪm tæg] *Namensschild*
administrative [əd'mɪnɪstrətɪv] *Verwaltungs-*
stuff [stʌf] *Kram*
register ['redʒɪstə] *zulassen*
number plate ['nʌmbə pleɪt] *Nummernschild*
sound [saʊnd] *klingen*
69 **receptionist** [rɪ'sepʃənɪst] *Empfang(smitarbeiter/in)*
all together [,ɔ:l tə'geðə] *insgesamt*
harbour ['hɑ:bə] *Hafen*
attract [ə'trækt] *anlocken, gewinnen*
patient ['peɪʃnt] *Patient/in*
security guard [sɪ'kjʊərəti gɑ:d] *Sicherheitsbeamtin/-er*
70 **pleased to meet you** [,pli:zd tə 'mi:t ju] *Es freut mich, Sie kennen zu lernen.*
cheerio [,tʃɪəri'əʊ] *tschüs, servus*
see you later [,si: ju 'leɪtə] *bis später*
71 **point to** ['pɔɪnt tə] *zeigen auf*
from the start [frɒm ðə 'stɑ:t] *von Anfang an*
rent [rent] *mieten*
staff [stɑ:f] *Personal, Mitarbeiter und Mitarbeiterinnen*
letting agency ['letɪŋ eɪdʒənsi] *(Immobilien-) Makler/in*
fortune ['fɔ:tʃu:n] *Vermögen*
boom [bu:m] *Hochkonjunktur*
share [ʃeə] *(die Wohnung mit jdm) teilen*
care [keə] *betreuen*
in return for [ɪn rɪ'tɜ:n fə] *als Gegenleistung für*
rubbish ['rʌbɪʃ] *Müll*
concentrate on ['kɒnsntreɪt ɒn] *sich konzentrieren auf*
72 **definition** [,defɪ'nɪʃn] *Definition*
special ['speʃl] *speziell*

friendly ['frendli] *freundlich, freundschaftlich*
slump [slʌmp] *Wirtschaftskrise*
recession [rɪ'seʃn] *Rezession*
outline ['aʊtlaɪn] *Kurzdarstellung, Skizze*
advantage [əd'vɑ:ntɪdʒ] *Vorteil, Vorzug*
pros [prəʊz] *Für, Pro*
disadvantage [ˌdɪsəd'vɑ:ntɪdʒ] *Nachteil*
cons [kɒnz] *Wider, Kontra*
opening ['əʊpnɪŋ] *Einleitung*
running cost [ˌrʌnɪŋ 'kɒst] *Unterhaltskosten, Betriebs(un)kosten*
chores [tʃɔ:z] *Routinearbeit, Arbeiten im Haushalt*
conclusion [kən'klu:ʒn] *(Schluss-)Folgerung, Entscheidung*
restate [ˌri:'steɪt] *neu formulieren, wiederholen*
reduce [rɪ'dju:s] *reduzieren, senken*
cost of living [ˌkɒst əv 'lɪvɪŋ] *Lebenshaltungskosten*
companionship [kəm'pæniənʃɪp] *Gesellschaft*
independence [ˌɪndɪ'pendəns] *Unabhängigkeit*
privacy ['prɪvəsi] *Privatsphäre*
consider [kən'sɪdə] *Rücksicht nehmen auf*
invite [ɪn'vaɪt] *einladen*
imagine [ɪ'mædʒɪn] *sich vorstellen*
value ['vælju:] *schätzen, Wert legen auf*
majority [mə'dʒɒrəti] *Mehrheit*
popular ['pɒpjələ] *beliebt*
73 **key word** ['ki: wɜ:d] *Schlüsselwort*
note down [ˌnəʊt 'daʊn] *aufschreiben, notieren*
whole [həʊl] *ganz*
arrange [ə'reɪndʒ] *anordnen*
systematically [ˌsɪstə'mætɪkli] *systematisch*
ecological [ˌi:kə'lɒdʒɪkl] *ökologisch*
writing ['raɪtɪŋ] *Schreiben*
otherwise ['ʌðəwaɪz] *sonst, ansonsten*
be able to [bi 'eɪbl tə] *können, in der Lage sein*
telephone ['telɪfəʊn] *telefonieren*
answer ['ɑ:nsə] *beantworten*
take notes [teɪk 'nəʊts] *(sich) Notizen machen*
used [ju:zd] *gebraucht*
motorcycle ['məʊtəsaɪkl] *Motorrad*
estate [ɪ'steɪt] *Kombi*
silver ['sɪlvə] *Silber*
mile [maɪl] *Meile*
MoT certificate [em əʊ ˌti: sə'tɪfɪkət] *TÜV-Bericht*
grab handle ['græb hændl] *Haltegriff*
perfect ['pɜ:fɪkt] *perfekt*
74 **holiday** ['hɒlədeɪ] *Feiertag*
borrow ['bɒrəʊ] *(sich) (aus)leihen*
bike [baɪk] *Fahrrad*
75 **scooter** ['sku:tə] *(Motor-)Roller*
connector [kə'nektə] *Bindewort*
furthermore [ˌfɜ:ðə'mɔ:] *ferner, zudem*
so that ['səʊ ðət] *damit*
to begin with [tə bɪ'gɪn wɪð] *zunächst*
rent [rent] *Miete*
completely [kəm'pli:tli] *völlig*

Unit 4

76 **carpenter** ['kɑ:pəntə] *Zimmermann, Tischler/in*
welder ['weldə] *Schweißer/in*
77 **worthwhile** [ˌwɜ:θ'waɪl] *lohnend*
future-orientated [ˌfju:tʃə 'ɔ:rienteɪtɪd] *zukunftsorientiert*
varied ['veərid] *abwechslungsreich*
secure [sɪ'kjuə] *sichern*
hard disk [ˌhɑ:d 'dɪsk] *Festplatte*
pipe wrench ['paɪp rentʃ] *Rohrzange*
ring spanner ['rɪŋ spænə] *Ringschlüssel*
test tube ['test tju:b] *Reagenzglas*
voltage ['vəʊltɪdʒ] *Spannung*
voltage tester ['vəʊltɪdʒ testə] *Spannungsprüfer*
welding torch ['weldɪŋ tɔ:tʃ] *Schweißbrenner*
tighten ['taɪtn] *anziehen, festschrauben*
loosen ['lu:sn] *lockern, lösen*
hexagonal [heks'ægənl] *sechseckig*
nut [nʌt] *Mutter*
bolt [bəʊlt] *Bolzen, Schraube*
cylindrical [sə'lɪndrɪkl] *zylindrisch*
design [dɪ'zaɪn] *entwerfen, entwickeln*
access ['ækses] *zugreifen auf, (Datei) öffnen*
welded ['weldɪd] *geschweißt*
joint [dʒɔɪnt] *Naht(stelle)*
amount [ə'maʊnt] *Menge*
chemical ['kemɪkl] *Chemikalie*
separate ['sepəreɪt] *trennen*
threaded collar [ˌθredɪd 'kɒlə] *Überwurfmutter*
78 **click** [klɪk] *(an)klicken*
call up [ˌkɔ:l 'ʌp] *aufrufen*
job description [dʒɒb dɪ'skrɪpʃn] *Stellenbeschreibung*
transmission [træns'mɪʃn] *Übertragung*
electricity [ɪˌlek'trɪsəti] *Elektrizität, Strom*
maintain [meɪn'teɪn] *warten*
electrically-operated [ɪˌlektrɪkli 'ɒpəreɪtɪd] *mit Elektroantrieb, elektrisch betrieben*
generator ['dʒenəreɪtə] *Generator*
heater ['hi:tə] *Heizgerät*
huge [hju:dʒ] *riesig*
appliance [ə'plaɪəns] *Gerät*
washing machine ['wɒʃɪŋ məʃi:n] *Waschmaschine*
ensure [ɪn'ʃʊə] *sicherstellen, gewährleisten*
supply [sə'plaɪ] *Lieferung, Zufuhr*
wiring ['waɪərɪŋ] *Verkabelung, Leitungen*
by law [baɪ 'lɔ:] *gesetzlich*
cooker ['kʊkə] *Herd, Kocher*
distribution [ˌdɪstrɪ'bju:ʃn] *Verteilung, Versorgung*
responsible [rɪ'spɒnsəbl] *verantwortlich*
erection [ɪ'rekʃn] *Errichtung*
pylon ['paɪlən] *(Strom-)Mast*
connection [kə'nekʃn] *Verbindung, Anschluss*
power line ['paʊə laɪn] *Stromkabel*
transformer substation [trænsˌfɔ:mə 'sʌbsteɪʃn] *Umspannwerk*
requirement [rɪ'kwaɪəmənt] *Anforderung, Voraussetzung*
in addition to [ɪn ə'dɪʃn tə] *zusätzlich (zu), neben*
required [rɪ'kwaɪəd] *erforderlich*
qualification [ˌkwɒlɪfɪ'keɪʃn] *Qualifikation, Fähigkeit*
dangerous ['deɪndʒərəs] *gefährlich*
remain [rɪ'meɪn] *bleiben*
calm [kɑ:m] *ruhig*
accurate ['ækjərət] *akkurat, genau*
under pressure [ˌʌndə 'preʃə] *unter Druck*
short cut ['ʃɔ:t kʌt] *Abkürzung*
physical ['fɪzɪkl] *körperlich, physisch*
eyesight ['aɪsaɪt] *Sehvermögen*

employment prospects [ɪm,plɔɪmənt 'prɒspekts] *Aussichten auf Arbeit, Beschäftigungschancen*
demand [dɪ'mɑ:nd] *Nachfrage, Bedarf*
as well as [əz 'wel əz] *sowie*
transport ['trænspɔ:t] *Verkehr, Transportwesen*
communications [kə,mju:nɪ'keɪʃnz] *Nachrichtenübermittlung*
79 **particularly** [pə'tɪkjələli] *besonders*
distribute [dɪ'strɪbju:t] *(aus)liefern, verteilen*
equip [ɪ'kwɪp] *ausrüsten, ausstatten*
generate ['dʒenəreɪt] *erzeugen*
heat [hi:t] *heizen*
qualify ['kwɒlɪfaɪ] *(sich) qualifizieren*
transform [træns'fɔ:m] *umwandeln*
transmit [træns'mɪt] *übertragen*
coal [kəʊl] *Kohle*
climate change ['klaɪmət tʃeɪndʒ] *Klimawandel*
personality [,pɜ:sə'næləti] *Persönlichkeit*
steam [sti:m] *Dampf*
discount ['dɪskaʊnt] *Rabatt, Skonto, Nachlass*
80 **employer** [ɪm'plɔɪə] *Arbeitgeber/in, Unternehmer/in*
questionnaire [,kwestʃə'neə] *Fragebogen*
what kind of [wɒt 'kaɪnd əv] *welche Art von*
honest ['ɒnɪst] *ehrlich*
add up [,æd 'ʌp] *zusammenzählen, addieren*
folded ['fəʊldɪd] *verschränkt*
clasped [klɑ:spt] *gefaltet*
hip [hɪp] *Hüfte*
touch [tʌtʃ] *anfassen, berühren*
ear [ɪə] *Ohr*
chin [tʃɪn] *Kinn*
hair [heə] *Haar/e*
knee [ni:] *Knie*
neat [ni:t] *ordentlich*
stretched out [,stretʃt 'aʊt] *ausgestreckt*
curled [kɜ:ld] *untergeschlagen*
beneath [bɪ'ni:θ] *unter(halb)*
funny ['fʌni] *lustig, witzig*
wide [waɪd] *weit*
laugh [lɑ:f] *lachen*
chuckle ['tʃʌkl] *in sich hineinlachen, kichern*
smile [smaɪl] *lächeln*
81 **notice** ['nəʊtɪs] *bemerken*
interrupt [,ɪntə'rʌpt] *unterbrechen*
extreme [ɪk'stri:m] *Extrem*
colour ['kʌlə] *Farbe*
purple ['pɜ:pl] *lila*
white [waɪt] *weiß*
brown [braʊn] *braun*
grey [greɪ] *grau*
go to sleep [gəʊ tə 'sli:p] *einschlafen*
lie [laɪ] *liegen*
stomach ['stʌmək] *Magen*
bent [bent] *gekrümmt, angewinkelt*
covers ['kʌvəz] *Bettdecke*
be asleep [bi ə'sli:p] *schlafen*
dream [dri:m] *Traum; träumen*
fight [faɪt] *kämpfen*
against [ə'genst] *gegen*
fly [flaɪ] *fliegen*
float [fləʊt] *schweben*
dreamless ['dri:mləs] *traumlos*
sweet [swi:t] *süß*
82 **scoring** ['skɔ:rɪŋ] *Punktestand, Auswertung*
dominant ['dɒmɪnənt] *beherrschend, dominant*
take control [teɪk kən'trəʊl] *die Kontrolle haben*
self-confident [,self 'kɒnfɪdənt] *selbstbewusst*
wish [wɪʃ] *wünschen*
exciting [ɪk'saɪtɪŋ] *aufregend*
risk-taker ['rɪsk teɪkə] *risikobereite Person*
colourful ['kʌləfl] *bunt, farbig*
impulsive [ɪm'pʌlsɪv] *impulsiv*
reckless ['rekləs] *unbekümmert, rücksichtslos*
certainly ['sɜ:tnli] *sicherlich*
have confidence in [həv 'kɒnfɪdəns ɪn] *Vertrauen haben zu*
judgement ['dʒʌdʒmənt] *Urteil*
fresh [freʃ] *frisch*
lively ['laɪvli] *lebendig*
charming ['tʃɑ:mɪŋ] *charmant*
amusing [ə'mju:zɪŋ] *amüsant*
practical ['præktɪkl] *praktisch (veranlagt)*
boring ['bɔ:rɪŋ] *langweilig*
centre of attention [,sentər əv ə'tenʃn] *Mittelpunkt der Aufmerksamkeit*
popularity [,pɒpju'lærəti] *Beliebtheit*
understanding [,ʌndə'stændɪŋ] *verständnisvoll*
sensible ['sensəbl] *vernünftig*
talented ['tæləntɪd] *begabt, talentiert*
modest ['mɒdɪst] *bescheiden*
make friends [meɪk 'frendz] *Freunde finden*
extremely [ɪk'stri:mli] *äußerst, außerordentlich*
loyal ['lɔɪəl] *treu, loyal*
circle ['sɜ:kl] *Kreis*
loyalty ['lɔɪəlti] *Treue, Loyalität*
painstaking ['peɪnzteɪkɪŋ] *sorgfältig, gewissenhaft*
cautious ['kɔ:ʃəs] *vorsichtig*
plodder ['plɒdə] *Arbeitstier*
examine [ɪg'zæmɪn] *untersuchen, prüfen*
shy [ʃaɪ] *schüchtern, scheu*
be afraid of [bi ə'freɪd əv] *sich fürchten vor*
upset [,ʌp'set] *verärgern, aufregen*
someone else [,sʌmwʌn 'els] *jemand anders*
decision [dɪ'sɪʒn] *Entscheidung*
worrier ['wʌriə] *sorgenvoller Mensch*
hidden depths [,hɪdn 'depθs] *verborgene Vorzüge*
handball coach ['hændbɔ:l kəʊtʃ] *Handballtrainer/in*
photographer [fə'tɒgrəfə] *Fotograf/in*
plumber ['plʌmə] *Installateur/in, Klempner/in*
racing driver ['reɪsɪŋ draɪvə] *Rennfahrer/in*
software engineer [,sɒftweə ,endʒɪ'nɪə] *Softwareentwickler/in*
soldier ['səʊldʒə] *Soldat/in*
83 **focus** ['fəʊkəs] *Brennpunkt, Schwerpunkt*
glass [glɑ:s] *Glas*
plastic ['plæstɪk] *Plastik, Kunststoff*
bottle ['bɒtl] *Flasche*
brainstorm ['breɪnstɔ:m] *(Ideen) sammeln (zu)*
compare [kəm'peə] *vergleichen*
aluminium [,æljə'mɪniəm] *Aluminium*
ceramic [sə'ræmɪk] *Keramik*
cotton ['kɒtn] *Baumwolle*
fibreglass ['faɪbəglɑ:s] *Glaswolle, -faser*
foam rubber [,fəʊm 'rʌbə] *Schaumgummi*
gold [gəʊld] *Gold*
leather ['leðə] *Leder*

rubber ['rʌbə] *Gummi*
steel [sti:l] *Stahl*
stone [stəʊn] *Stein*
wax [wæks] *Wachs*
wool [wʊl] *Wolle*
composite ['kɒmpəzɪt] *zusammengesetzt*
natural ['nætʃrəl] *natürlich*
process ['prəʊses] *bearbeiten*
synthetic [sɪn'θetɪk] *synthetisch*
assembly line [ə'sembli laɪn] *Montage-, Fließband*
more or less [,mɔ:r ɔ: 'les] *mehr oder weniger*
raw [rɔ:] *Roh(material)*
occur [ə'kɜ:] *vorkommen*
man-made ['mæn meɪd] *künstlich, Kunst-*
84 **polyester paint** [,pɒli'estə peɪnt] *Polyesterfarbe*
brass [brɑ:s] *Messing*
PVC [,pi: vi: 'si:] *PVC, Polyvinylchlorid*
petrol ['petrəl] *Benzin*
nylon ['naɪlɒn] *Nylon*
piston ['pɪstən] *Kolben*
property ['prɒpəti] *Eigenschaft*
brittle ['brɪtl] *brüchig, spröde*
ductile ['dʌktaɪl] *(ver)formbar, ausziehbar*
durable ['djʊərəbl] *haltbar, widerstandsfähig, langlebig*
impact-resistant [,ɪmpækt rɪ'zɪstənt] *schlagfest*
non-corrosive [,nɒn kə'rəʊsɪv] *nicht rostend*
transparent [træns'pærənt] *durchsichtig, durchlässig*
bend [bend] *(sich) biegen (lassen)*
break [breɪk] *(zer)brechen*
crack [kræk] *knacken, brechen, springen*
implant ['ɪmplɑ:nt] *Implantat*
exercise book ['eksəsaɪz bʊk] *(Schul-)Heft*
stainless steel [,steɪnləs 'sti:l] *rostfreier Stahl*
nickel ['nɪkl] *Nickel*
titanium [tɪ'teɪniəm] *Titan*
85 **pinning** ['pɪnɪŋ] *Nageln*
bonding ['bɒndɪŋ] *Verklebung*
rivet ['rɪvɪt] *Niete*
staple ['steɪpl] *Heftklammer*
clip [klɪp] *Klammer, Flansch, Lasche, Schelle, Halterung*
chemical fixing agent [,kemɪkl 'fɪksɪŋ eɪdʒənt] *chemischer Klebstoff*
adhesive [əd'hi:sɪv] *Klebstoff*
heavy-duty [,hevi 'dju:ti] *strapazierfähig*
chair [tʃeə] *Stuhl*
glue [glu:] *Klebstoff, Leim*
peg [peg] *Haken*
air filter ['eə fɪltə] *Luftfilter*
beam [bi:m] *Balken*
section ['sekʃn] *Teil, Abschnitt*
tiler ['taɪlə] *Fliesenleger/in*
lay [leɪ] *verlegen*
tile [taɪl] *Kachel, Fliese*
lamp-holder ['læmp həʊldə] *Lampenfassung*
ceiling ['si:lɪŋ] *(Zimmer-)Decke*
riveter ['rɪvɪtə] *Nieter/in*
steel plate ['sti:l pleɪt] *Stahlplatte*
86 **hydroelectric** [,haɪdrəʊɪ'lektrɪk] *hydroelektrisch, Wasserkraft-*
drill bit ['drɪl bɪt] *Bohrer*
mineral water ['mɪnərəl wɔ:tə] *Mineralwasser*
countable ['kaʊntəbl] *zählbar*
uncountable [,ʌn'kaʊntəbl] *nicht zählbar*
divide into [dɪ'vaɪd ɪntə] *unterteilen in*
87 **dynamo** ['daɪnəməʊ] *Lichtmaschine, Dynamo*
evidence ['evɪdəns] *Beweismaterial, -mittel*
crowd [kraʊd] *(Menschen-)Menge*
base [beɪs] *Grund-*
hurry up! [,hʌri 'ʌp] *Beeilung!Beeil dich!*
wages ['weɪdʒɪz] *Löhne*
lay off [,leɪ 'ɒf] *(zeitweilig) entlassen*
skier ['ski:ə] *Skiläufer/in*
slope [sləʊp] *(Ab-)Hang*

Unit 5

88 **in advance** [ɪn əd'vɑ:ns] *im Voraus, vorher*
aim [eɪm] *Ziel, Absicht*
guess [ges] *(er)raten*
delay [dɪ'leɪ] *Verspätung, Verzögerung*
overcrowding [,əʊvə'kraʊdɪŋ] *Überfüllung*
lead to ['li:d tə] *führen zu*
drop [drɒp] *Rückgang*
business travel ['bɪznəs trævl] *Geschäftsreisen*
upside ['ʌpsaɪd] *Vorzug*
difference ['dɪfrəns] *Unterschied*
89 **checklist** ['tʃeklɪst] *Checkliste*
complaint [kəm'pleɪnt] *Beanstandung, Reklamation*
review [rɪ'vju:] *überprüfen*
draw up [,drɔ: 'ʌp] *aufstellen*
participant [pɑ:'tɪsɪpənt] *Teilnehmer/in*
finalise ['faɪnəlaɪz] *(endgültig) festlegen*
agenda [ə'dʒendə] *Tagesordnung*
update [,ʌp'deɪt] *aktualisieren, auf den neuesten Stand bringen*
prepare [prɪ'peə] *vorbereiten*
documentation [,dɒkjumen'teɪʃn] *Dokumentation, Unterlagen*
organise ['ɔ:gənaɪz] *organisieren*
refreshments [rɪ'freʃmənts] *Erfrischungen, Getränke*
back to back [,bæk tə 'bæk] *Rücken an Rücken*
telephone call ['telɪfəʊn kɔ:l] *Telefongespräch*
90 **evaluate** [ɪ'væljueɪt] *aus-, bewerten*
venue ['venju:] *(Tagungs-)Ort*
Customer Services [,kʌstəmə 'sɜ:vɪsɪz] *Kundendienst*
minutes ['mɪnɪts] *Protokoll*
topic ['tɒpɪk] *Thema*
greeting ['gri:tɪŋ] *Begrüßung*
introduction [,ɪntrə'dʌkʃn] *Einführung*
current ['kʌrənt] *derzeitig, aktuell, gegenwärtig*
discussion [dɪ'skʌʃn] *Diskussion, Erörterung*
action ['ækʃn] *Vorgehen*
any other business [,eni ,ʌðə 'bɪznəs] *Verschiedenes*
91 **speaker** ['spi:kə] *Redner/in, Person, die spricht*
point of view [,pɔɪnt əv 'vju:] *Standpunkt, Ansicht*
point [pɔɪnt] *Punkt, Thema*
expert ['ekspɜ:t] *Fachmann/frau, Experte/Expertin*
solve [sɒlv] *lösen*
latest ['leɪtɪst] *neueste, letzte*
session ['seʃn] *Sitzung*
call a meeting [,kɔ:l ə 'mi:tɪŋ] *eine Sitzung anberaumen*
92 **shape** [ʃeɪp] *Form*

plane [pleɪn] *flach, eben*
solid ['sɒlɪd] *fest, dreidimensional*
dimension [dɪ'menʃn] *Abmessung, Maß*
width [wɪdθ] *Breite*
fall into ['fɔ:l ɪntə] *sich unterteilen lassen*
rectilinear [ˌrektɪ'lɪniə] *gradlinig*
curvilinear [ˌkɜ:vɪ'lɪniə] *gekrümmt*
angle ['æŋgl] *Winkel*
curved [kɜ:vd] *gekrümmt*
combination [ˌkɒmbɪ'neɪʃn] *Kombination, Verbindung*
isosceles [ai'sɒsəli:z] *gleichschenklig*
triangle ['traɪæŋgl] *Dreieck*
degree [dɪ'gri:] *Grad*
apex ['eɪpeks] *Spitze, Scheitel*
combine [kəm'baɪn] *zusammenfügen, kombinieren*
sided ['saɪdɪd] *-seitig*
square [skweə] *Quadrat*
equal ['i:kwəl] *gleich*
rectangle ['rektæŋgl] *Rechteck*
parallel ['pærəlel] *parallel*
93 **pentagon** ['pentəgən] *Fünfeck*
hexagon ['heksəgən] *Sechseck*
octagon ['ɒktəgən] *Achteck*
polygon ['pɒlɪgən] *Vieleck, Polygon*
regular ['regjələ] *regelmäßig*
equilateral triangle [ˌi:kwɪ'lætərəl 'traɪæŋgl] *gleichseitiges Dreieck*
irregular [ɪ'regjələ] *unregelmäßig*
rhomboid ['rɒmbɔɪd] *Rhomboid*
other than ['ʌðə ðən] *außer*
right angle ['raɪt æŋgl] *rechter Winkel*
acute [ə'kju:t] *spitz*
obtuse [əb'tju:s] *stumpf*
straight line [streɪt 'laɪn] *Gerade*
reflex ['ri:fleks] *überstumpf*
94 **measurement** ['meʒəmənt] *Abmessung, Maß*
reverse [rɪ'vɜ:s] *umgekehrt*
alphabetical [ˌælfə'betɪkl] *alphabetisch*
depth [depθ] *Tiefe*
semicircle ['semisɜ:kl] *Halbkreis*
ellipse [ɪ'lɪps] *Ellipse*
ridge tent ['rɪdʒ tent] *Hauszelt*
wheel [wi:l] *Rad*
sugar cube ['ʃʊgə kju:b] *Würfelzucker*
clock-face ['klɒk feɪs] *Zifferblatt*
three-dimensional [ˌθri: dɪ'menʃənl] *dreidimensional*
apart [ə'pɑ:t] *getrennt*
95 **pencil** ['pensl] *Bleistift*
slimline TV ['slɪmlaɪn ti: vi:] *Flachbildschirm*
reverse [rɪ'vɜ:s] *umkehren, vertauschen*
96 **uncle** ['ʌŋkl] *Onkel*
contract ['kɒntrækt] *Vertrag, Auftrag*
build [bɪld] *bauen, errichten*
stadium ['steɪdiəm] *Stadion*
certain ['sɜ:tn] *sicher*
Romania [ru:'meɪniə] *Rumänien*
spot check ['spɒt tʃek] *Stichprobe*
cancel ['kænsl] *absagen*
lifestyle ['laɪfstaɪl] *Lebensstil*
lose weight [lu:z 'weɪt] *abnehmen*
take up [ˌteɪk 'ʌp] *anfangen mit*
fast food [ˌfɑ:st 'fu:d] *Schnellgerichte*
healthy ['helθi] *gesund*
lift [lɪft] *Aufzug, Fahrstuhl*
97 **campsite** ['kæmpsaɪt] *Campingplatz*
first-aid kit [ˌfɜ:st 'eɪd kɪt] *Verbandskasten*
photocopy ['fəʊtəʊkɒpi] *fotokopieren*
reserve [rɪ'zɜ:v] *reservieren, buchen*
cloud [klaʊd] *Wolke*
watch out! [ˌwɒtʃ 'aʊt] *Vorsicht! Pass auf!*
pallet ['pælət] *Palette*
forecast ['fɔ:kɑ:st] *(Wetter-)Vorhersage*
trade fair ['treɪd feə] *Handelsmesse*

Unit 6

98 **assignment** [ə'saɪnmənt] *Aufgabe*
bottom-hung window [ˌbɒtəm hʌŋ 'wɪndəʊ] *Kippfenster*
combination boiler [ˌkɒmbɪ'neɪʃn bɔɪlə] *Wasserkocher*
dishwasher ['dɪʃwɒʃə] *Geschirrspüler*
outside blinds [ˌaʊtsaɪd 'blaɪndz] *Rolläden*
cistern ['sɪstən] *Spülkasten*
washbasin ['wɒʃbeɪsn] *Waschbecken*
plughole ['plughəʊl] *Abfluss*
household appliance [ˌhaʊshəʊld ə'plaɪəns] *Haushaltsgerät*
survey ['sɜ:veɪ] *(Meinungs-)Umfrage*
explanation [ˌeksplə'neɪʃn] *Erklärung*
keyed [ki:d] *mit Schlüssel*
chuck [tʃʌk] *Bohrfutter*
anticlockwise [ˌænti'klɒkwaɪz] *gegen den Uhrzeigersinn*
jaw [dʒɔ:] *Bohrfutter*
sense of rotation [ˌsens əv rəʊ'teɪʃn] *Drehrichtung*
clockwise ['klɒkwaɪz] *im Uhrzeigersinn*
counterclockwise [ˌkaʊntə'klɒkwaɪz] *gegen den Uhrzeigersinn, links herum*
slider switch ['slaɪdə swɪtʃ] *Schiebeschalter*
vice versa [ˌvaɪs 'vɜ:sə] *(und) umgekehrt*
mode [məʊd] *Modus, Art*
setting ring ['setɪŋ rɪŋ] *Einstellring*
numbered ['nʌmbəd] *nummeriert*
torque [tɔ:k] *Drehkraft, Drehmoment*
size [saɪz] *Größe*
insert [ɪn'sɜ:t] *einsetzen*
keyless ['ki:ləs] *ohne Schlüssel*
outer ['aʊtə] *äußere, Außen-*
collar ['kɒlə] *Ring*
unnumbered [ˌʌn'nʌmbəd] *unnummeriert*
marker ['mɑ:kə] *Markierung*
arrowhead ['ærəʊhed] *Pfeilspitze*
99 **heat pump** ['hi:t pʌmp] *Wärmepumpe*
unlike [ˌʌn'laɪk] *anders als, im Gegensatz zu*
gas-fired [ˌgæs'faɪəd] *Gasheiz(kessel)*
burn [bɜ:n] *verbrennen*
consume [kən'sju:m] *verbrauchen*
characteristic [ˌkærəktə'rɪstɪk] *Eigenschaft*
exceptionally [ɪk'sepʃənəli] *außerordentlich*
technically ['teknɪkli] *technisch*
function ['fʌŋkʃn] *funktionieren, arbeiten*
principle ['prɪnsəpl] *Grundsatz, Prinzip*
refrigerator [rɪ'frɪdʒəreɪtə] *Kühlschrank*
air conditioner ['eə kəndɪʃnə] *Klimaanlage*
liquid ['lɪkwɪd] *flüssig; Flüssigkeit*
absorb [əb'sɔ:b] *aufnehmen, absorbieren*

release [rɪ'liːs] *freisetzen, -geben*
return [rɪ'tɜːn] *zurückkehren*
state [steɪt] *Zustand*
refrigerant [rɪ'frɪdʒərənt] *Kühlflüssigkeit*
pump [pʌmp] *pumpen*
coil [kɔɪl] *Schlange*
tubing ['tjuːbɪŋ] *Röhren*
gaseous ['gæsiəs] *gasförmig*
evaporator coil [ɪ'væpəreɪtə kɔɪl] *Verdampferschlange*
cool [kuːl] *kühl*
duct [dʌkt] *Rohr*
meanwhile ['miːnwaɪl] *inzwischen*
compressor [kəm'presə] *Kompressor*
liquify ['lɪkwɪfaɪ] *flüsssig werden, (sich) verflüssigen*
condensor [kən'densə] *Kondensator*
compressed [kəm'prest] *komprimiert*
atmosphere ['ætməsfɪə] *Atmosphäre*
extract [ɪk'strækt] *gewinnen, entziehen*
range [reɪndʒ] *Bereich*
progressively [prə'gresɪvli] *zunehmend*
drop [drɒp] *fallen*
significant [sɪg'nɪfɪkənt] *spürbar*
supplement ['sʌplɪment] *ergänzen*
auxiliary [ɔːg'zɪliəri] *zusätzlich*
solid-fuel [ˌsɒlɪd 'fjuːəl] *Brennstoff-*
stove [stəʊv] *Ofen, Herd*
practicable ['præktɪkəbl] *brauchbar*
region ['riːdʒən] *Gegend*
northern ['nɔːðən] *nördlich, Nord-*
solution [sə'luːʃn] *Lösung*
southern ['sʌðən] *Süd-, südlich*
ambient ['æmbiənt] *Umgebungs-*
100 **explain** [ɪk'spleɪn] *erklären, erläutern*
environmentally friendly [ɪnˌvaɪrənˌmentəli 'frendli] *umweltfreundlich*
point out [ˌpɔɪnt 'aʊt] *erläutern, erklären*
similarity [ˌsɪmə'lærəti] *Ähnlichkeit*
Sweden ['swiːdn] *Schweden*
presentation [ˌprezn'teɪʃn] *Vorstellung, Präsentation*
convenience [kən'viːniəns] *Komfort, Bequemlichkeit*
ease [iːz] *Leichtigkeit*
operation [ˌɒpə'reɪʃn] *Betrieb, Bedienung*
environmental [ɪnˌvaɪrən'mentl] *Umwelt-*
benefit ['benɪfɪt] *Vorteil, Nutzen*
effective [ɪ'fektɪv] *effektiv, wirkungsvoll*
script [skrɪpt] *Manuskript*
prompt card ['prɒmpt kɑːd] *Stichwortkarte*
key point ['kiː pɔɪnt] *Kernpunkt*
scan [skæn] *überfliegen*
statue ['stætʃuː] *Statue, Standbild*
vary ['veəri] *variieren, ändern*
volume ['vɒljuːm] *Lautstärke*
voice [vɔɪs] *Stimme*
get across [ˌget ə'krɒs] *vermitteln*
shout [ʃaʊt] *schreien*
emphasise ['emfəsaɪz] *betonen, hervorheben*
gesture ['dʒestʃə] *Geste*
overdo it [ˌəʊvə'duː ɪt] *übertreiben*
authorised dealer [ˌɔːθəraɪzd 'diːlə] *Vertragshändler*
versatile ['vɜːsətaɪl] *vielseitig*
outstanding [aʊt'stændɪŋ] *hervorstechend*
feature ['fiːtʃə] *Merkmal*
occupier ['ɒkjupaɪə] *Bewohner/in*
override [ˌəʊvə'raɪd] *Eingriff (in Automatik)*
101 **underfloor** [ˌʌndə'flɔː] *Fußboden-*
due to ['djuː tə] *wegen, aufgrund von*
radiation [ˌreɪdi'eɪʃn] *(Ab-)Strahlung, Abgabe*
essential [ɪ'senʃl] *(absolut) notwendig*
considerably [kən'sɪdərəbli] *beträchtlich*
conventional [kən'venʃənl] *herkömmlich, konventionell*
loss [lɒs] *Verlust*
comfort ['kʌmfət] *Komfort*
moreover [mɔːr'əʊvə] *außerdem, darüber hinaus*
circulation [ˌsɜːkjə'leɪʃn] *Zirkulation*
dust [dʌst] *Staub*
dust mite ['dʌst maɪt] *(Haus-)Staubmilbe*
allergic [ə'lɜːdʒɪk] *allergisch*
constant ['kɒnstənt] *konstant, gleich bleibend*
guide [gaɪd] *leiten*
totally ['təʊtəli] *völlig, total*
cover ['kʌvə] *bedecken, einschließen*
screed [skriːd] *Estrich*
ecologically neutral [ˌiːkəˌlɒdʒɪkli 'njuːtrəl] *ökologisch neutral*
CFC [ˌsiː ef 'siː] *FCKW*
recyclable [ˌriː'saɪkləbl] *wieder verwertbar, recycelbar*
dry [draɪ] *trocken*
strip [strɪp] *Streifen*
plate [pleɪt] *Platte*
foil [fɔɪl] *Folie*
humidity [hjuː'mɪdəti] *Feuchtigkeit*
loaded ['ləʊdɪd] *beladen*
chlorofluorocarbon [ˌklɔːrəʊˌflʊərəʊ'kɑːbən] *Fluorchlorkohlenwasserstoff*
102 **mono plate** ['mɒnəʊ pleɪt] *einteilige Fläche*
additional [ə'dɪʃənl] *zusätzlich*
unheated [ʌn'hiːtɪd] *ungeheizt*
soil [sɔɪl] *(Erd-)Boden*
plaster ['plɑːstə] *Gips, Verputz*
conduction element [kən'dʌkʃn elɪmənt] *Leitungselement*
covering ['kʌvərɪŋ] *Belag, Verkleidung*
elastic [ɪ'læstɪk] *elastisch*
skirting board ['skɜːtɪŋ bɔːd] *Fußleiste*
sealing ['siːlɪŋ] *Versiegelung*
layer ['leɪə] *Schicht*
residual [rɪ'zɪdjuəl] *Rest-, restlich*
moisture ['mɔɪstʃə] *Feuchtigkeit*
103 **relatively** ['relətɪvli] *relativ, verhältnismäßig*
effect [ɪ'fekt] *(Aus-)Wirkung, Einfluss*
zone [zəʊn] *Zone, Gebiet*
structure ['strʌktʃə] *Struktur, Aufbau*
fitted carpet [ˌfɪtɪd 'kɑːpɪt] *Teppichboden*
first floor [ˌfɜːst 'flɔː] *erste Etage, erster Stock*
cellar ['selə] *Keller*
recessed [rɪ'sest] *eingelassen, umbaut*
balcony ['bælkəni] *Balkon*
detailed ['diːteɪld] *detailliert, ausführlich*
include [ɪn'kluːd] *einbeziehen, angeben*
104 **flooring** ['flɔːrɪŋ] *Bodenbelag*
invest [ɪn'vest] *investieren*
rebuild [ˌriː'bɪld] *wieder aufbauen*

renovation [ˌrenəˈveɪʃn] *Renovierung*
wooden [ˈwʊdn] *hölzern, Holz-*
consignment [kənˈsaɪnmənt] *(Waren-)Sendung, Lieferung*
test [test] *überprüfen*
specified [ˈspesɪfaɪd] *angegeben*
evidently [ˈevɪdəntli] *offensichtlich*
satisfied [ˈsætɪsfaɪd] *zufrieden*
105 **fax machine** [ˈfæks məʃiːn] *Faxgerät*
switch on [ˌswɪtʃ ˈɒn] *einschalten*
weekly [ˈwiːkli] *wöchentlich*
personnel [ˌpɜːsəˈnel] *Personal, Mitarbeiter und Mitarbeiterinnen*
construction site [kənˈstrʌkʃn saɪt] *Baustelle*
seal [siːl] *Siegel*
obtain [əbˈteɪn] *erhalten, bekommen*
emergency [ɪˈmɜːdʒənsi] *Notfall*

Unit 7

106 **bicycle** [ˈbaɪsɪkl] *Fahrrad*
horse [hɔːs] *Pferd*
rowing boat [ˈrəʊɪŋ bəʊt] *Ruderboot*
tram [træm] *Straßenbahn*
means of transport [miːnz əf ˈtrænspɔːt] *Verkehrsmittel*
107 **fuel cell** [ˈfjuːəl sel] *Brennstoffzelle*
hybrid [ˈhaɪbrɪd] *hybrid*
hydrogen [ˈhaɪdrədʒən] *Wasserstoff*
luxury [ˈlʌkʃəri] *Luxus*
sedan [sɪˈdæn] *Limousine*
make sure [ˌmeɪk ˈʃʊə] *sich vergewissern, dafür sorgen*
conversion [kənˈvɜːʃn] *Umwandlung*
run down [ˌrʌn ˈdaʊn] *sich entladen*
compete [kəmˈpiːt] *konkurrieren*
direct current [dəˌrekt ˈkʌrənt] *Gleichstrom*
precious metal [ˌpreʃəs ˈmetl] *Edelmetall*
catalyst [ˈkætəlɪst] *Katalysator*
deteriorate [dɪˈtɪəriəreɪt] *(sich) verschlechtern*
volatile [ˈvɒlətaɪl] *flüchtig*
108 **electrochemical** [ɪˌlektrəʊˈkemɪkl] *elektrochemisch*
energy [ˈenədʒi] *Energie*
oxygen [ˈɒksɪdʒən] *Sauerstoff*
feed into [ˈfiːd ɪntə] *einspeisen*
develop [dɪˈveləp] *entwickeln*
internal combustion engine [ɪnˌtɜːnl kəmˈbʌstʃən endʒɪn] *Verbrennungsmotor*
notebook [ˈnəʊtbʊk] *Laptop*
harmful [ˈhɑːmfl] *schädlich*
emissions [ɪˈmɪʃnz] *Abgase*
toothbrush [ˈtuːθbrʌʃ] *Zahnbürste*
it is no surprise that [ɪts nəʊ səˈpraɪs ðət] *es ist kein Wunder, dass*
electrolyte [ɪˈlektrəlaɪt] *Elektrolyt*
portable [ˈpɔːtəbl] *transportabel, tragbar*
however [haʊˈevə] *doch, dennoch*
platinum [ˈplætɪnəm] *Platin*
secondly [ˈsekəndli] *zweitens*
frequent [ˈfriːkwənt] *häufig, oft*
turn off [ˌtɜːn ˈɒf] *abschalten, abstellen*
pure [pjʊə] *rein*
danger [ˈdeɪndʒə] *Gefahr*
explosion [ɪkˈspləʊʒn] *Explosion*
purpose [ˈpɜːpəs] *Zweck*
109 **proton exchange membrane** [ˌprəʊtɒn ɪksˌtʃeɪndʒ ˈmembreɪn] *Protonaustauschmembran*
enter [ˈentə] *eindringen (in)*
chemical reaction [ˌkemɪkl riˈækʃn] *chemische Reaktion*
porous [ˈpɔːrəs] *porös*
electron [ɪˈlektrɒn] *Elektron*
DC [ˌdiː ˈsiː] *Gleichstrom*
ion [ˈaɪən] *Ion*
electrical charge [ɪˌlektrɪkl ˈtʃɑːdʒ] *elektrische Ladung*
load [ləʊd] *Ladung*
excess [ˈekses] *überschüssig, überzählig*
output [ˈaʊtpʊt] *Leistung*
increase [ɪnˈkriːs] *steigern, erhöhen*
limitless [ˈlɪmɪtləs] *unbegrenzt, grenzenlos*
contaminated [kənˈtæmɪneɪtɪd] *verunreinigt*
purified [ˈpjʊərɪfaɪd] *gereinigt*
impracticable [ɪmˈpræktɪkəbl] *undurchführbar*
pollution [pəˈluːʃn] *(Umwelt-, Luft-)Verschmutzung*
110 **superior** [suːˈpɪəriə] *überlegen*
firstly [ˈfɜːstli] *zuerst*
whereas [ˌweərˈæz] *während*
kinetic [kɪˈnetɪk] *kinetisch*
movement [ˈmuːvmənt] *Bewegung*
disaster [dɪˈzɑːstə] *Katastrophe, Unglück*
agreed [əˈgriːd] *übereinstimmend(e Meinung)*
argument [ˈɑːgjumənt] *Argument, Einwand*
traditional [trəˈdɪʃənl] *traditionell*
recharge [ˌriːˈtʃɑːdʒ] *aufladen*
legendary [ˈledʒəndri] *legendär*
strict [strɪkt] *streng*
ahead of its time [əˌhed əv ɪts ˈtaɪm] *seiner Zeit voraus*
top speed [ˌtɒp ˈspiːd] *Höchstgeschwindigkeit*
acceleration [əkˌseləˈreɪʃn] *Beschleunigung*
force [fɔːs] *zwingen*
advance [ədˈvɑːns] *Fortschritt*
nickel-metal hydride [ˌnɪkl metl ˈhaɪdraɪd] *Nickel-Metall-Hydrid*
lead [led] *Blei*
acid [ˈæsɪd] *Säure*
double [ˈdʌbl] *verdoppeln*
in the case of [ɪn ðə ˈkeɪs əv] *im Falle von*
out of date [ˌaʊt əv ˈdeɪt] *veraltet, überholt*
numerous [ˈnjuːmərəs] *zahlreich*
111 **alternative** [ɔːlˈtɜːnətɪv] *Alternative*
hybrid [ˈhaɪbrɪd] *Hybridfahrzeug*
waste [weɪst] *vergeuden, verschwenden*
controversy [ˈkɒntrəvɜːsi] *Kontroverse*
source [sɔːs] *Quelle*
dependency on [dɪˈpendənsi ɒn] *Abhängigkeit von*
import [ɪmˈpɔːt] *importieren*
overcome [ˌəʊvəˈkʌm] *überwinden*
rewrite [ˌriːˈraɪt] *umschreiben*
air pollution [ˌeə pəˈluːʃn] *Luftverschmutzung*
since then [sɪns ˈðen] *seitdem*
microfilter [ˈmaɪkrəʊfɪltə] *Mikrofilter*
criticise [ˈkrɪtɪsaɪz] *kritisieren*
abolish [əˈbɒlɪʃ] *abschaffen, streichen*
sales tax [ˈseɪlz tæks] *Mehrwerts-, Umsatzsteuer*
low-energy light bulb [ləʊ ˌenədʒi ˈlaɪt bʌlb] *Sparlampe*

rechargeable [ˌriːˈtʃɑːdʒəbl] *aufladbar*
112 **Channel tunnel** [ˈtʃænl tʌnl] *Kanaltunnel*
multi-purpose [ˌmʌlti ˈpɜːpəs] *Mehrzweck-*
fire engine [ˈfaɪər endʒɪn] *Feuerwehrauto*
put out [ˌpʊt ˈaʊt] *löschen*
injure [ˈɪndʒə] *verletzen*
113 **consist of** [kənˈsɪst əv] *bestehen aus*
enable [ɪˈneɪbl] *(es) ermöglichen*
pulse [pʌls] *pulsieren (lassen)*
in succession [ɪn səkˈseʃn] *nacheinander*
controller [kənˈtrəʊlə] *Steuergerät*
pedal [ˈpedl] *Pedal*
potentiometer [pəˌtenʃiˈɒmɪtə] *Potentiometer*
rate [reɪt] *Geschwindigkeit*
signal [ˈsɪgnəl] *Signal, Zeichen*
power unit [ˈpaʊə juːnɪt] *Antriebsaggregat*
array [əˈreɪ] *Ansammlung, Reihe*
in all [ɪn ˈɔːl] *insgesamt, alles in allem*
accelerator pedal [əkˈseləreɪtə pedl] *Gaspedal*
variable resistor [ˌveəriəbl rɪˈzɪstə] *regelbarer Widerstand*
in between [ɪn bɪˈtwiːn] *dazwischen*
flow [fləʊ] *fließen*
in order to [ɪn ˈɔːdə tə] *um zu*
halfway [ˌhɑːfˈweɪ] *halb*
pulsing [ˈpʌlsɪŋ] *Pulsieren*
114 **AC** [ˌeɪ ˈsiː] *Wechselstrom*
phase [feɪz] *Phase*
provide [prəˈvaɪd] *liefern*
charger [ˈtʃɑːdʒə] *Aufladevorrichtung*
accessory [əkˈsesəri] *Zusatz*
on the one hand [ɒn ðə ˈwʌn hænd] *einerseits*
refuel [ˌriːˈfjuːəl] *auftanken*
weight-carrier [ˈweɪt kæriə] *Lastenträger*
consumption [kənˈsʌmpʃn] *Verbrauch*
narrow [ˈnærəʊ] *eingeschränkt*
uncomfortable [ʌnˈkʌmftəbl] *unbequem*
adult [ˈædʌlt] *Erwachsene/r*
115 **caller** [ˈkɔːlə] *Anrufer/in*
owner [ˈəʊnə] *Besitzer/in*
doubt [daʊt] *bezweifeln*
get rid of [get ˈrɪd əv] *abschaffen*
wind farm [ˈwɪnd fɑːm] *Windkraftanlage*
solar [ˈsəʊlə] *Sonnen-, Solar-*
radical [ˈrædɪkl] *radikal*
policy [ˈpɒləsi] *Politik*
destroy [dɪˈstrɔɪ] *zerstören, vernichten*
countryside [ˈkʌntrisaɪd] *Landschaft*
carbon [ˈkɑːbən] *Kohlenstoff, -dioxid*
be worried [bi ˈwʌrid] *sich Sorgen machen*
ferrous [ˈferəs] *Eisen-*
fluid [ˈfluːɪd] *Flüssigkeit*
light van [ˌlaɪt ˈvæn] *leichter Lieferwagen*
population [ˌpɒpjuˈleɪʃn] *Bevölkerung*
as a whole [əz ə ˈhəʊl] *insgesamt*
net [net] *netto*
take-back service [ˌteɪk ˈbæk sɜːvɪs] *Rücknahmedienst*
pie-chart [ˈpaɪ tʃɑːt] *Torten-, Kreisdiagramm*
disposal [dɪˈspəʊzl] *Entsorgung*
116 **powerful** [ˈpaʊəfl] *mächtig, stark*
go well [ˌgəʊ ˈwel] *gut (aus)gehen*
litre [ˈliːtə] *Liter*
recent [ˈriːsnt] *jüngst, aktuell*
absurd [əbˈsɜːd] *absurd*
credit crunch [ˈkredɪt krʌntʃ] *Kreditknappheit, -klemme*
slight [slaɪt] *gering, leicht*
sharp [ʃɑːp] *scharf, stark*
surprising [səˈpraɪzɪŋ] *überraschend, erstaunlich*
rough [rʌf] *grob, hart*
motoring [ˈməʊtərɪŋ] *Autofahren*
get from A to B [get frəm ˌeɪ tə ˈbiː] *von A nach B gelangen*
thrill [θrɪl] *(aufregendes) Gefühl*
terrific [təˈrɪfɪk] *großartig, phantastisch*
absolutely [ˈæbsəluːtli] *absolut*
official [əˈfɪʃl] *offiziell*
open-top [ˌəʊpən ˈtɒp] *mit offenem Verdeck*
sports car [ˈspɔːts kɑː] *Sportwagen*
redesigned [ˌriːdɪˈzaɪnd] *neu konstruiert*
authentic [ɔːˈθentɪk] *authentisch*
robust [rəʊˈbʌst] *robust*
handling [ˈhændlɪŋ] *Handhabung*
proud [praʊd] *stolz*
117 **comparison** [kəmˈpærɪsn] *Vergleich, Steigerung*
reliability [rɪˌlaɪəˈbɪləti] *Zuverlässigkeit*
capacity [kəˈpæsəti] *Kapazität, Fassungsvermögen*
in comparison with [ɪn kəmˈpærɪsn wɪð] *im Vergleich zu*
on the other hand [ɒn ði ˈʌðə hænd] *andererseits*
reliable [rɪˈlaɪəbl] *zuverlässig*
user manual [ˌjuːzə ˈmænjuəl] *Benutzerhandbuch*

Unit 8

118 **planet** [ˈplænɪt] *Planet*
climate [ˈklaɪmət] *Klima*
expert opinion [ˌekspɜːt əˈpɪnjən] *Expertenmeinung*
by far [baɪ ˈfɑː] *bei weitem*
threat [θret] *Bedrohung*
global warming [ˌgləʊbl ˈwɔːmɪŋ] *Erderwärmung, Treibhauseffekt*
flood [flʌd] *Flut, Überschwemmung*
rise [raɪz] *(an)steigen*
water level [ˈwɔːtə ˌlevl] *Wasserspiegel*
drown [draʊn] *ertrinken, ertränken*
low-lying [ˈləʊ laɪɪŋ] *niedrig liegend*
community [kəˈmjuːnəti] *Gemeinde*
glacier [ˈglæsiə] *Gletscher*
ice-cap [ˈaɪs kæp] *Eiskappe*
melt [melt] *schmelzen*
drought [draʊt] *Trockenzeit, Dürre(periode)*
forest [ˈfɒrɪst] *Wald*
devastate [ˈdevəsteɪt] *zerstören, verwüsten*
farmland [ˈfɑːmlænd] *landwirtschaftlich genutzte Fläche*
hurricane [ˈhʌrɪkən] *Orkan, Wirbelsturm*
severe [sɪˈvɪə] *schwer, stark*
storm [stɔːm] *Unwetter, Sturm*
destructive [dɪˈstrʌktɪv] *zerstörerisch*
species [ˈspiːʃiːz] *Art*
pest [pest] *Plage, Schädling*
worrying [ˈwʌriɪŋ] *beunruhigend*
119 **take on** [ˈteɪk ɒn] *Einstellung zu*
opinion survey [əˌpɪnjən ˈsɜːveɪ] *Meinungsumfrage*
reality [riˈæləti] *Wirklichkeit, Realität*
human [ˈhjuːmən] *menschlich*

activity [æk'tɪvəti] *Aktivität, Tätigkeit*
minority [maɪ'nɒrəti] *Minderheit*
largely ['lɑːdʒli] *zum größten Teil, weit gehend*
fluctuation [ˌflʌktʃu'eɪʃn] *Schwankung*
media hype ['miːdiə haɪp] *Medienrummel*
observer [əb'zɜːvə] *Beobachter/in*
winner ['wɪnə] *Gewinner/in*
phenomenon [fə'nɒmɪnən] *Phänomen*
aware [ə'weə] *bewusst, aufmerksam*
the rich [ðə 'rɪtʃ] *die Reichen*
exaggerated [ɪg'zædʒəreɪtɪd] *übertrieben*
submit [səb'mɪt] *übersenden, aufgeben*
compile [kəm'paɪl] *zusammenstellen*
as far as ... goes [əz ˌfɑːr əz ˌgəʊz] *was ... anbetrifft*
either way [ˌaɪðə 'weɪ] *in jeder/keiner Hinsicht*
loser ['luːzə] *Verlierer/in*
120 **debate** [dɪ'beɪt] *Debatte*
correct [kə'rekt] *korrigieren*
renewable [rɪ'njuːəbl] *erneuerbar*
flood defences [ˌflʌd dɪ'fensɪz] *Dämme*
kid oneself [kɪd] *sich etw vormachen*
consequence ['kɒnsɪkwəns] *Folge, Konsequenz*
behaviour [bɪ'heɪvjə] *Verhalten, Benehmen*
sector ['sektə] *Sektor, Bereich*
emit [ɪ'mɪt] *emittieren, ausstoßen*
agriculture ['ægrɪkʌltʃə] *Landwirtschaft*
frequency ['friːkwənsi] *Häufigkeit*
bar chart ['bɑː tʃɑːt] *Säulen-, Balkendiagramm*
billion ['bɪliən] *Milliarde*
burning ['bɜːnɪŋ] *Verbrennung, Brand*
deforestation [diːˌfɒrɪ'steɪʃn] *Abholzung, Entwaldung*
environmental pollution [ɪnˌvaɪrənˌmentl pə'luːʃn] *Umweltverschmutzung*
line graph ['laɪn grɑːf] *(Kurven-, Linien-)Diagramm*
fossil fuel ['fɒsl fjuːəl] *fossiler Brennstoff*
121 **relate to** [rɪ'leɪt tə] *beziehen auf*
tropical ['trɒpɪkl] *tropisch, Tropen-*
horizontal [ˌhɒrɪ'zɒntl] *horizontal, waagerecht*
axis ['æksɪs] *Achse*
tonne [tʌn] *Tonne*
present [prɪ'zent] *vorstellen, präsentieren*
basically ['beɪsɪkli] *grundsätzlich*
deal with ['diːl wɪð] *sich befassen mit, behandeln*
relationship [rɪ'leɪʃnʃɪp] *Beziehung, Verhältnis*
steady ['stedi] *stetig, regel-, gleichmäßig*
rise [raɪz] *Anstieg, Erhöhung*
decrease ['diːkriːs] *Rückgang, Abnahme*
climb [klaɪm] *klettern, ansteigen*
gradually ['grædʒuəli] *allmählich, schrittweise*
steep [stiːp] *steil*
peak [piːk] *gipfeln*
reach a peak [ˌriːtʃ ə 'piːk] *einen Höchststand erreichen*
bottom out [ˌbɒtəm 'aʊt] *die Talsohle erreichen*
minimum ['mɪnɪməm] *Mindest-*
segment ['segmənt] *Teilstück, Segment*
point in time [pɔɪnt ɪn 'taɪm] *Zeitpunkt*
in relation to [ɪn rɪ'leɪʃn tə] *in Bezug auf*
122 **escape** [ɪ'skeɪp] *entweichen*
report back [rɪˌpɔːt 'bæk] *Bericht erstatten, berichten über*
ban [bæn] *verbieten, verzichten auf*
standby switch ['stændbaɪ swɪtʃ] *Ruhezustandsschalter*
nuclear ['njuːkliə] *Atom-, Kern-*
aviation [ˌeɪvi'eɪʃn] *Luftfahrt, Flug-*
subsidise ['sʌbsɪdaɪz] *subventionieren*
inland waterway [ˌɪnlənd 'wɔːtəweɪ] *Binnenschifffahrtsstraße*
barge [bɑːdʒ] *Last-, Binnenschiff*
cycle track ['saɪkl træk] *Fahrradweg*
suggestion [sə'dʒestʃən] *Vorschlag*
suggest [sə'dʒest] *vorschlagen*
disagree [ˌdɪsə'griː] *anderer Meinung sein, widersprechen, nicht zustimmen*
123 **poll** [pəʊl] *Umfrage*
respondent [rɪ'spɒndənt] *Befragte/r*
promising ['prɒmɪsɪŋ] *vielversprechend*
awareness [ə'weənəs] *Bewusstsein*
gasoline ['gæsəliːn] *Benzin*
none [nʌn] *kein*
serious ['sɪəriəs] *ernst(haft)*
federal ['fedərəl] *Bundes-*
124 **geothermal** [ˌdʒiːəʊ'θɜːml] *Erdwärme-*
carbon-intensive [ˌkɑːbən ɪn'tensɪv] *kohlenstoffintensiv*
wise [waɪz] *weise, klug*
close to home [kləʊs tə 'həʊm] *nahe am Haus*
interior [ɪn'tɪəriə] *Innere(s)*
core [kɔː] *Kern*
decay [dɪ'keɪ] *Zerfall*
radioactive [ˌreɪdiəʊ'æktɪv] *radioaktiv*
particle ['pɑːtɪkl] *Teilchen*
earth [ɜːθ] *Erde*
mass [mæs] *Masse*
molten ['məʊltən] *flüssig*
diameter [daɪ'æmɪtə] *Durchmesser*
mantle ['mæntl] *Ummantelung*
crust [krʌst] *Kruste*
ocean ['əʊʃn] *Ozean, Meer*
continental plate [ˌkɒntɪnentl 'pleɪt] *Kontinentalplatte*
earthquake ['ɜːθkweɪk] *Erdbeben*
volcano [vɒl'keɪnəʊ] *Vulkan*
lava ['lɑːvə] *Lava*
erupt [ɪ'rʌpt] *ausbrechen*
partly ['pɑːtli] *teilweise, zum Teil*
deep [diːp] *tief*
underground ['ʌndəgraʊnd] *unterirdisch*
around the world [əraʊnd ðə 'wɜːld] *auf der ganzen Welt*
dig [dɪg] *graben*
well [wel] *Brunnen*
visible ['vɪzəbl] *sichtbar*
above ground [ə'bʌv graʊnd] *oberirdisch*
fumarole ['fjuːmərəʊl] *vulkanische Dampfquelle*
volcanic [vɒl'kænɪk] *vulkanisch*
hot spring [ˌhɒt 'sprɪŋ] *heiße Quelle*
geyser ['giːzə] *Geysir*
Iceland ['aɪslənd] *Island*
accessible [ək'sesəbl] *zugänglich*
boundary ['baʊndri] *Grenze*
edge [edʒ] *Rand*
Pacific Ocean [pəˌsɪfɪk 'əʊʃn] *Pazifik, Pazifischer Ozean*
125 **trap** [træp] *einschließen, speichern*
fractured ['fræktʃəd] *brüchig*
fault [fɔːlt] *(Verwerfungs-)Spalte*
reservoir ['rezəvwɑː] *Stausee, -becken*

geologist [dʒi'ɒlədʒɪst] *Geologe/Geologin*
drill [drɪl] *bohren*
exist [ɪg'zɪst] *existieren*
matter ['mætə] *Materie*
skin [skɪn] *Haut*
tiny ['taɪni] *winzig*
126 **paragraph** ['pærəgrɑ:f] *Absatz, (kurzer) Artikel*
binary ['baɪnəri] *binär*
determine [dɪ'tɜ:mɪn] *bestimmen*
emerge [ɪ'mɜ:dʒ] *austreten*
harness ['hɑ:nɪs] *nutzbar machen*
in this way [ɪn ðɪs 'weɪ] *auf diese Art und Weise*
flash steam plant [ˌflæʃ sti:m 'plɑ:nt] *Heiß-dampfkraftwerk*
spray [spreɪ] *sprühen*
instantly ['ɪnstəntli] *sofort*
vaporise ['veɪpəraɪz] *zerstäuben*
vapour ['veɪpə] *Dampf*
turbine ['tɜ:baɪn] *Turbine*
differ ['dɪfə] *sich unterscheiden*
127 **relevant** ['reləvənt] *entsprechend*
secondary ['sekəndri] *sekundär, Neben-*
hence [hens] *daher*
boiling point ['bɔɪlɪŋ pɔɪnt] *Siedepunkt*
crisis ['kraɪsɪs] *Krise*
appoint [ə'pɔɪnt] *ernennen*
findings ['faɪndɪŋz] *Ergebnisse*
view [vju:] *Ansicht, Standpunkt*
wherever [weər'evə] *wo (immer)*
make use of [meɪk 'ju:s əv] *benutzen*
audience-friendly ['ɔ:diəns frendli] *publikums-freundlich*
shortage ['ʃɔ:tɪdʒ] *Mangel*
Brazil [brə'zɪl] *Brasilien*
among other things [əˌmʌŋ 'ʌðə θɪŋz] *unter anderem*
civil ['sɪvl] *zivil*
unrest [ʌn'rest] *Unruhe(n)*
transformation [ˌtrænzfə'meɪʃn] *Umwandlung*
replacement [rɪ'pleɪsmənt] *Ersatz*
conservation [ˌkɒnsə'veɪʃn] *Erhaltung*
vs. ['vɜ:səs] *gegen(über)*
beautiful ['bju:tɪfl] *schön*
recommendation [ˌrekəmen'deɪʃn] *Empfehlung*
trust [trʌst] *Stiftung, Trust*
128 **pipeline** ['paɪplaɪn] *Rohrleitung, Pipeline*
town centre [ˌtaʊn 'sentə] *Stadtzentrum, City*
dirty ['dɜ:ti] *schmutzig*
close down [ˌkləʊz 'daʊn] *schließen, abschalten*
public transport [ˌpʌblɪk 'trænspɔ:t] *öffentliche Verkehrsmittel*
even if ['i:vn ɪf] *selbst wenn*
take off [ˌteɪk 'ɒf] *starten, (ab)fliegen*
win [wɪn] *gewinnen*
arrival [ə'raɪvl] *Ankunft*
129 **economical** [ˌi:kə'nɒmɪkl] *sparsam, wirtschaft-lich*
evening class ['i:vnɪŋ klɑ:s] *Abendschule*
lever ['li:və] *Hebel*
automatically [ˌɔ:tə'mætɪkli] *automatisch*
downtime ['daʊntaɪm] *Ausfallzeit*
earn [ɜ:n] *verdienen*
favourite ['feɪvərɪt] *Lieblings-*
attack [ə'tæk] *angreifen*

Unit 9

130 **CCTV camera** [ˌsi: si: ti: 'vi: 'kæmərə] *Über-wachungskamera*
closed-circuit television [ˌkləʊzd, sɜ:kɪt 'telɪvɪʒn] *Überwachungskamera(s)*
telephone tapping ['telɪfəʊn tæpɪŋ] *Telefon-überwachung*
biometric ID card [ˌbaɪəʊˌmetrɪk ˌaɪ 'di: kɑ:d] *bio-metrischer Ausweis*
tracking ['trækɪŋ] *Zurückverfolgen*
GPS tracking system [ˌdʒi: pi: 'es 'trækɪŋ sɪstəm] *GPS Satellitennavigationssystem*
global positioning system [ˌgləʊbl pə'zɪʃənɪŋ sɪstəm] *GPS*
database ['deɪtəbeɪs] *Datenbank*
security firm [sɪ'kjʊərəti fɜ:m] *Sicherheitsfirma, Wachdienst*
informer [ɪn'fɔ:mə] *Denunziant/in*
identity [aɪ'dentəti] *Identität*
whether ['weðə] *ob*
getaway car ['getəweɪ kɑ:] *Fluchtwagen*
bank robbery ['bæŋk rɒbəri] *Bankraub*
statistics [stə'tɪstɪks] *Statistiken, statistische Angaben*
neighbour ['neɪbə] *Nachbar/in*
cruel [kru:əl] *grausam*
acceptable [ək'septəbl] *akzeptabel*
131 **infrastructure** ['ɪnfrəstrʌktʃə] *Infrastruktur*
railway station ['reɪlweɪ steɪʃn] *Bahnhof*
swivel ['swɪvl] *(sich) drehen*
increasingly [ɪn'kri:sɪŋli] *zunehmend*
wireless ['waɪələs] *drahtlos, Funk-*
cabling ['keɪblɪŋ] *Verkabelung*
suspend [sə'spend] *hängen*
mount [maʊnt] *montieren, anbringen*
pole [pəʊl] *Pfahl, Mast*
wrongdoer ['rɒŋdu:ə] *Missetäter/in*
frighten away [ˌfraɪnt ə'weɪ] *abschrecken, verscheuchen*
under surveillance [ˌʌndə sɜ:'veɪləns] *unter Beobachtung*
department store [dɪ'pɑ:tmənt stɔ:] *Kauf-, Warenhaus*
continuous [kən'tɪnjuəs] *ständig*
exclusive [ɪk'sklu:sɪv] *exklusiv*
upmarket [ˌʌp'mɑ:kɪt] *oberes Marktsegment*
fashion ['fæʃn] *Mode*
boutique [bu:'ti:k] *Boutique*
jeweller ['dʒu:ələ] *Juwelier/in*
high-class [ˌhaɪ 'klɑ:s] *erstklassig, Nobel-*
lobby ['lɒbi] *Foyer, Eingangshalle*
discreet [dɪ'skri:t] *diskret*
micro-camera [ˌmaɪkrəʊ 'kæmərə] *Mikrokamera*
light fitting ['laɪt fɪtɪŋ] *Lampenfassung*
spy on ['spaɪ ɒn] *ausspionieren*
132 **specification** [ˌspesɪfɪ'keɪʃn] *Angabe*
storage ['stɔ:rɪdʒ] *Speicherung, Archivierung*
backup ['bækʌp] *Sicherungskopie*
compression [kəm'preʃn] *Kompression*
viewing screen ['vju:ɪŋ skri:n] *Bildschirm*
network ['netwɜ:k] *Netz(werk)*
record [rɪ'kɔ:d] *aufnehmen, -zeichnen*
playback ['pleɪbæk] *Playback*
bay [beɪ] *Ausgang, Anschluss*
unlimited [ʌn'lɪmɪtɪd] *unbegrenzt*

download [ˌdaʊn'ləʊd] *herunterladen*
VCR [ˌviː siː 'ɑː] *Videorekorder*
input ['ɪnpʊt] *Eingabe*
channel ['tʃænl] *Kanal*
voltage controlled amplifier (VCA) [ˌvəʊltɪdʒ kənˌtrəʊld 'æmplɪfaɪə] *VCA-Verstärker*
loop [luːp] *Schleife*
video graphics array (VGA) ['vɪdiəʊ ˌgræfɪks ə'reɪ] *Videofilmaufzeichnung*
optional ['ɒpʃənl] *wahlweise*
enhanced [ɪn'hɑːnst] *verbessert*
motion activated [ˌməʊʃn 'æktɪveɪtɪd] *durch Bewegung aktiviert*
frames per second (FPS) [ˌfreɪmz pə 'sekənd] *Bilder pro Sekunde*
picture in picture (PIP) [ˌpɪktʃə ɪn 'pɪktʃə] *Bild im Bild*
transmission control protocol (TCP) [trænsˌmɪʃn kənˌtrəʊl 'prəʊtəkɒl] *Netzwerkprotokoll im Internet*
viewer ['vjuːə] *Überwachung*
power supply ['paʊə səplaɪ] *Stromanschluss*
snake [sneɪk] *Schlange*
Ltd ['lɪmɪtɪd] *(Gesellschaft) mit beschränkter Haftung, GmbH*
enquiry [ɪn'kwaɪəri] *Anfrage*
package ['pækɪdʒ] *Paket*
recommend [ˌrekə'mend] *empfehlen*
comprise [kəm'praɪz] *umfassen*
133 **like the look of** [laɪk ðə 'lʊk əv] *gefallen das Aussehen*
formal ['fɔːml] *formal, förmlich*
business premises [ˌbɪznəs 'premɪsɪz] *Geschäftsräume, Firmengelände*
list price ['lɪst praɪs] *Listenpreis*
on receipt of [ɒn rɪ'siːt əv] *nach Eingang*
buyer ['baɪə] *Käufer/in*
134 **filling station** ['fɪlɪŋ steɪʃn] *Tankstelle*
vending machine ['vendɪŋ məʃiːn] *Verkaufsautomat*
spirits ['spɪrɪts] *Spirituosen*
freezer ['friːzə] *Gefriertruhe, Tiefkühltruhe*
magazine [ˌmægə'ziːn] *Zeitschrift*
car wash ['kɑː wɒʃ] *Autowaschanlage*
perimeter fence [pə'rɪmɪtə fens] *Umzäunung*
cashier [kæ'ʃɪə] *Kassierer/in*
position [pə'zɪʃn] *anbringen*
135 **presenter** [prɪ'zentə] *Moderator/in*
criminal lawyer ['krɪmɪnl lɔːjə] *Anwalt für Strafsachen*
specialise in ['speʃəlaɪz ɪn] *sich spezialisieren auf*
civil rights [ˌsɪvl 'raɪts] *Bürgerrechte*
human rights lawyer [ˌhjuːmən 'raɪts lɔːjə] *Anwalt für Menschenrechte*
commercial lawyer [kə'mɜːʃl lɔːjə] *Wirtschaftsanwalt*
takeover ['teɪkəʊvə] *Übernahme, Fusion*
street crime ['striːt kraɪm] *Straßenkriminalität*
communication [kəˌmjuːnɪ'keɪʃn] *Kommunikation*
credit card ['kredɪt kɑːd] *Kreditkarte*
with great difficulty [wɪð greɪt 'dɪfɪkəlti] *unter großen Schwierigkeiten*
photograph ['fəʊtəgrɑːf] *fotografieren*
letter to the editor [ˌletə tə ði 'edɪtə] *Leserbrief*
and so on [ənd 'səʊ ɒn] *und so weiter*
glad [glæd] *froh*
convenient [kən'viːniənt] *bequem*
retailer ['riːteɪlə] *Einzelhändler/in*
interest ['ɪntrəst] *interessieren*
hysterical [hɪ'sterɪkl] *hysterisch*
feeling ['fiːlɪŋ] *Gefühl*
encyclopedia [ɪnˌsaɪklə'piːdiə] *Enzyklopädie, Lexikon*
dictionary ['dɪkʃənri] *Wörterbuch*
postcard ['pəʊstkɑːd] *Postkarte*
136 **electronic article surveillance** [ɪˌlektrɒnɪk ˌɑːtɪkl sɜː'veɪləns] *elektronische Artikelüberwachung*
shoplifting ['ʃɒplɪftɪŋ] *Ladendiebstahl*
tolerant ['tɒlərənt] *tolerant*
push up [ˌpʊʃ 'ʌp] *hochtreiben*
pleasant ['pleznt] *angenehm*
shopkeeper ['ʃɒpkiːpə] *Ladeninhaber/in, -besitzer/in*
lock away [lɒk ə'weɪ] *wegschließen*
out of reach [aʊt əv 'riːtʃ] *außer Reichweite*
unpopular [ʌn'pɒpjələ] *unbeliebt, unpopulär*
sales assistant ['seɪlz əsɪstənt] *Verkäufer/in*
try on [ˌtraɪ 'ɒn] *anprobieren*
store detective [ˌstɔː dɪ'tektɪv] *Warenhausdetektiv/in*
anti-shoplifting [ˌænti 'ʃɒplɪftɪŋ] *gegen Ladendiebstahl*
adopt [ə'dɒpt] *einführen*
tagging ['tægɪŋ] *Diebstahlsicherung*
practically ['præktɪkli] *praktisch*
clothing ['kləʊðɪŋ] *Bekleidung, Kleider*
amazing [ə'meɪzɪŋ] *erstaunlich*
label ['leɪbl] *Schild(chen), Etikett*
shop assistant ['ʃɒp əsɪstənt] *Verkäufer/in*
deactivate [ˌdiː'æktɪveɪt] *deaktivieren*
swipe [swaɪp] *darüber ziehen*
handheld ['hændheld] *Hand-(Gerät)*
scanner ['skænə] *Scanner, Lesegerät*
radio frequency [ˌreɪdiəʊ 'friːkwənsi] *Rundfunkfrequenz*
have sth against [həv ə'genst] *etw dagegen haben*
tend to ['tend tə] *dazu neigen*
tagged [tægd] *gegen Diebstahl gesichert*
137 **adviser** [əd'vaɪzə] *Berater/in*
surveillance equipment [sɜːˌveɪləns ɪ'kwɪpmənt] *Überwachungsgerät(e)*
interference [ˌɪntə'fɪərəns] *Störung*
receiver [rɪ'siːvə] *Empfänger*
shoulder height ['ʃəʊldə haɪt] *Schulterhöhe*
transmitter [træns'mɪtə] *Sender*
removable [rɪ'muːvəbl] *abnehmbar*
reuse [ˌriː'juːz] *wieder verwenden*
pedestal ['pedɪstl] *Sockel, Fundament*
antenna [æn'tenə] *Antenne*
preset [ˌpriː'set] *voreingestellt*
handle ['hændl] *umgehen mit, erfassen*
miniature ['mɪnətʃə] *Miniatur-*
respond [rɪ'spɒnd] *reagieren*
good heavens [gʊd 'hevnz] *ach du meine Güte*
138 **consumer** [kən'sjuːmə] *Verbraucher/in*
theft [θeft] *Diebstahl*
developed world [dɪˌveləpt 'wɜːld] *Industrieländer*
PIN code ['pɪn kəʊd] *PIN-Nummer*
violence ['vaɪələns] *Gewalt*

worldwide ['wɜ:ldwaɪd] *weltweit*
International Mobile station Equipment Identity (IMEI) [ˌɪntənæʃnəl ˌməʊbaɪl steɪʃn ɪˌkwɪpmənt aɪ'dentəti] *internationale 15-stellige Seriennummer für Handys*
unique [ju'ni:k] *einmalig*
databank ['deɪtəbæŋk] *Datenbank*
reprogram [ˌri:'prəʊgræm] *neu programmieren*
thus [ðʌs] *so (mit)*
useless ['ju:sləs] *nutzlos, unbrauchbar*
forum ['fɔ:rəm] *Forum*
ineffectiveness [ˌɪnɪ'fektɪvnəs] *Wirkungslosigkeit*
cellphone ['selfəʊn] *Handy*
139 **admit** [əd'mɪt] *zugeben, gestehen*
fail [feɪl] *scheitern, (es) nicht schaffen*
whatever [wɒt'evə] *was auch immer*
housing estate ['haʊzɪŋ ɪsteɪt] *Wohnsiedlung, -anlage*
refuse [rɪ'fju:z] *(es) ablehnen, sich weigern*
business people ['bɪznəs pi:pl] *Geschäftsleute*
borough ['bʌrə] *Stadtbezirk*
recognise ['rekəgnaɪz] *erkennen*
face [feɪs] *Gesicht*
hooligan ['hu:lɪgən] *Randalierer/in, Rowdy*
thug [θʌg] *Schläger*
put up [ˌpʊt 'ʌp] *anbringen*
main road [meɪn 'rəʊd] *Hauptstraße*
ever since [ˌevə 'sɪns] *seitdem*
succeed in [sək'si:d ɪn] *es erreichen*
dislike [dɪs'laɪk] *nicht mögen*
under the eye of [ˌʌndə ði 'aɪ əv] *unter dem Objektiv von*
set up [ˌset 'ʌp] *aufstellen*

Unit 10

140 **nanotech** ['nænəʊtek] *Nanotechnik*
ingredient [ɪn'gri:diənt] *Bestandteil*
bond [bɒnd] *verbinden, verkleben*
nanoengine ['nænəʊendʒɪn] *Nanomotor*
listener ['lɪsnə] *Zuhörer/in*
explore [ɪk'splɔ:] *erforschen, erkunden*
headline ['hedlaɪn] *Schlagzeile, Überschrift*
importance [ɪm'pɔ:tns] *Wichtigkeit, Bedeutung*
brainstorming ['breɪnstɔ:mɪŋ] *Ideensammeln*
specific [spə'sɪfɪk] *speziell, bestimmt*
141 **screw thread contact** [ˌskru: θred 'kɒntækt] *Gewindekontakt*
filament ['fɪləmənt] *(Glüh-)Faden*
inert gas [ɪ'nɜ:t gæs] *Edelgas*
mount [maʊnt] *Sockel*
conductor [kən'dʌktə] *Leiter*
foot contact ['fʊt kɒntækt] *Fußkontakt*
incandescent [ˌɪnkæn'desnt] *weiß glühend*
vertical ['vɜ:tɪkl] *senkrecht, vertikal*
house [haʊz] *unterbringen*
sealed [si:ld] *versiegelt, abgedichtet*
inefficient [ˌɪnɪ'fɪʃnt] *ineffizient, unwirtschaftlich*
dissipate ['dɪsɪpeɪt] *ableiten*
cooling fan ['ku:lɪŋ fæn] *Kühllüfter*
crazy ['kreɪzi] *verrückt*
phase out [ˌfeɪz 'aʊt] *auslaufen lassen*
saving ['seɪvɪŋ] *Einsparung, Ersparnis*
halogen bulb ['hælədʒən bʌlb] *Halogenlampe*
fluorescent [flɔ:'resnt] *Leucht-, fluoreszierend*
ecologically friendly [ˌi:kə'lɒdʒɪkli frendli] *umweltfreundlich*
toxic ['tɒksɪk] *giftig, toxisch*
mercury ['mɜ:kjəri] *Quecksilber*
142 **daylight** ['deɪlaɪt] *Tageslicht*
give rise to [gɪv 'raɪz tə] *veranlassen, erregen*
in favour of [ɪn 'feɪvər əv] *zugunsten von*
death [deθ] *Tod*
famous ['feɪməs] *berühmt*
sadly ['sædli] *traurigerweise*
accidental [ˌæksɪ'dentl] *zufällig*
discovery [dɪ'skʌvəri] *Entdeckung*
in common [ɪn 'kɒmən] *gemeinsam*
illuminated [ɪ'lu:mɪneɪtɪd] *beleuchtet*
Christmas tree ['krɪsməs tri:] *Weihnachtsbaum*
candle ['kændl] *Kerze*
graduate ['grædʒuət] *Hochschulabsolvent/in*
quantum dot ['kwɒntəm dɒt] *Quantenpunkt*
crystal ['krɪstl] *Kristall*
contain [kən'teɪn] *enthalten*
excited [ɪk'saɪtɪd] *erregt*
bundle ['bʌndl] *Bündel, Paket*
batch [bætʃ] *Gruppe*
subjected to [səb'dʒektɪd tə] *ausgesetzt*
electric charge [ɪˌlektrɪk 'tʃɑ:dʒ] *elektrische Ladung*
bright [braɪt] *hell, strahlend*
vivid ['vɪvɪd] *lebhaft*
laser beam ['leɪzə bi:m] *Laserstrahl*
unexpected [ˌʌnɪk'spektɪd] *unerwartet*
glow [gləʊ] *Leuchten, Schein*
be supposed to [bi sə'pəʊzd tə] *sollen*
coat [kəʊt] *beschichten*
lumpy ['lʌmpi] *klumpig*
pretty ['prɪti] *schön, nett*
143 **yellowish-white** [ˌjeləʊɪʃ 'waɪt] *gelblich-weiß*
breakthrough ['breɪkθru:] *Durchbruch*
adapt [ə'dæpt] *adaptieren*
whitish ['waɪtɪʃ] *weißlich*
hue [hju:] *Farbton*
improvement [ɪm'pru:vmənt] *Verbesserung*
ordinary ['ɔ:dnri] *gewöhnlich, normal*
estimate ['estɪmeɪt] *schätzen*
series production [ˌsɪəri:z prə'dʌkʃn] *Serienproduktion, -fertigung*
paint [peɪnt] *malen*
just about [dʒʌst ə'baʊt] *fast*
rainbow ['reɪnbəʊ] *Regenbogen*
pick out [ˌpɪk 'aʊt] *auswählen, heraussuchen*
expose to [ɪk'spəʊz tə] *aussetzen*
studio apartment [ˌstju:diəʊ ə'pɑ:tmənt] *Einzimmerwohnung*
144 **home improvement** [həʊm ɪm'pru:vmənt] *Renovierung*
chair [tʃeə] *leiten*
central heating programmer [ˌsentrəl hi:tɪŋ 'prəʊgræmə] *Zentralheizungsregler*
radiator thermostat [ˌreɪdieɪtə 'θɜ:məstæt] *Heizkörperthermostat*

condensing boiler [kənˌdensɪŋ 'bɔɪlə] *Kondenskessel*
145 **price increase** ['praɪs ɪnkri:s] *Preiserhöhung*
clothes [kləʊðz] *Kleidung*
fix [fɪks] *Lösung*

houseowner ['haʊsəʊnə] *Hausbesitzer/in*
heat-reactive ['hi:t ri:æktɪv] *auf Hitze reagierend*
technological [ˌteknə'lɒdʒɪkl] *technologisch*
householder ['haʊshəʊldə] *Hauseigentümer/in*
heat-sensitive ['hi:t sensətɪv] *hitzeempfindlich*
guest [gest] *Gast*
insulating material [ˌɪnsjuleɪtɪŋ mə'tɪəriəl] *Isoliermaterial*
146 **assist** [ə'sɪst] *helfen, unterstützen*
fully-wired [ˌfʊli 'waɪəd] *voll verkabelt*
high-tech [ˌhaɪ 'tek] *hoch technisiert*
wonder ['wʌndə] *Wunder*
labour-saving ['leɪbə seɪvɪŋ] *arbeitssparend*
gadget ['gædʒɪt] *Vorrichtung, Gerät*
so-called [ˌsəʊ 'kɔ:ld] *so genannt*
smart [smɑ:t] *intelligent*
threaten ['θretn] *bedrohen*
emphasis ['emfəsɪs] *Betonung*
butt [bʌt] *Fass, Tonne*
rainwater ['reɪnwɔ:tə] *Regenwasser*
filter ['fɪltə] *filtern*
greywater ['greɪwɔ:tə] *Brauchwasser*
flush [flʌʃ] *spülen*
dependence [dɪ'pendəns] *Abhängigkeit*
solar cell ['səʊlə sel] *Solarzelle*
voltaic [vɒl'teɪɪk] *voltaisch*
roof [ru:f] *Dach*
wind turbine ['wɪnd tɜ:baɪn] *Windturbine*
triple-glazed ['trɪpl gleɪzd] *dreifach verglast*
conserve [kən'sɜ:v] *erhalten*
147 **airtight** ['eətaɪt] *luftdicht*
frame [freɪm] *Rahmen*
ventilation [ˌventɪ'leɪʃn] *Lüftung, Belüftung*
photovoltaic [ˌfəʊtəʊvɒl'teɪɪk] *photoelektrisch*
total-wrap insulation [ˌtəʊtl ræp ˌɪnsju'leɪʃn] *Rundumisolierung*
148 **introductory** [ˌɪntrə'dʌktəri] *Einführungs-, einleitend*
flight [flaɪt] *Flug*
solar panel [ˌsəʊlə 'pænl] *Sonnenkollektor*
dental appointment [ˌdentl ə'pɔɪntmənt] *Zahnarzttermin*
expenses [ɪk'spensɪz] *Spesen(abrechnung)*
printer ['prɪntə] *Drucker/in*
catalogue ['kætəlɒg] *Katalog*
149 **reported speech** [rɪˌpɔ:tɪd 'spi:tʃ] *indirekte Rede*
cupboard ['kʌbəd] *Schrank*
ear muffs ['ɪə mʌfs] *Ohrenschützer*
keep away from [ˌki:p ə'weɪ frəm] *sich fernhalten von*
jewellery ['dʒu:əlri] *Schmuck*
loose [lu:s] *weit (geschnitten), locker*

Diese Liste enthält alle Wörter in alphabetischer Reihenfolge.
Nicht aufgeführt sind die ca. 500 vorausgesetzten Wörter.
Die Zahl nach dem Stichwort bezieht sich auf die Seite, auf der das Wort zum ersten Mal erscheint.

A

able to, be ~ *73* *können, in der Lage sein*
a pair of *16* *ein Paar, eine (Zange)*
abbreviation *66* *Abkürzung*
abolish *111* *abschaffen, streichen*
above ground *124* *oberirdisch*
abroad *46* *im Ausland*
absolutely *116* *absolut*
absorb *99* *aufnehmen, absorbieren*
absurd *116* *absurd*
AC *114* *Wechselstrom*
acceleration *110* *Beschleunigung*
accelerator pedal *113* *Gaspedal*
acceptable *130* *akzeptabel*
access *77* *zugreifen auf, (Datei) öffnen*
access ramp *60* *Zufahrtsrampe*
accessible *124* *zugänglich*
accessory *114* *Zusatz*
accident *28* *Unfall*
accidental *142* *zufällig*
accommodation *67* *Unterkunft*
according to *65* *nach Aussage von*
account *9* *Konto*
accurate *78* *akkurat, genau*
acid *110* *Säure*
across *42* *(quer) über*
act out *63* *vorspielen, aufführen*
action *90* *Vorgehen*
activity *119* *Aktivität, Tätigkeit*
acute *93* *spitz*
ad *46* *Anzeige*
adapt *143* *adaptieren*
add up *80* *zusammenzählen, addieren*
additional *102* *zusätzlich*
address *49* *Adresse*
adhesive *85* *Klebstoff*
adjust *64* *einstellen*
adjustable *16* *verstellbar*
administration *40* *Verwaltung*
administrative *68* *Verwaltungs-*
admit *139* *zugeben, gestehen*
adopt *136* *einführen*
adult *114* *Erwachsene/r*
advance *110* *Fortschritt*
advanced *47* *fortgeschritten*
advantage *72* *Vorteil, Vorzug*
advertisement *48* *Anzeige*
adviser *137* *Berater/in*
Afghan *6* *Afghane/Afghanin; afghanisch*
Afghanistan *6* *Afghanistan*
afraid of, be ~ *82* *sich fürchten vor*
after all *60* *schließlich*
after-sales *41* *Kundendienst-*
against *81* *gegen*
agenda *89* *Tagesordnung*
agree *52* *übereinstimmen*
agreed *110* *übereinstimmend(e Meinung)*
agriculture *120* *Landwirtschaft*
ahead of its time *110* *seiner Zeit voraus*
aim *88* *Ziel, Absicht*
air conditioner *99* *Klimaanlage*
air filter *85* *Luftfilter*
air pollution *111* *Luftverschmutzung*
aircraft *20* *Flugzeug*
airtight *147* *luftdicht*
Albania *6* *Albanien*
Albanian *6* *Albanier/in; albanisch*
alcohol *21* *Alkohol*
alcoholic *21* *alkoholisch*
all modern conveniences *67* *aller Komfort*
all over the world *23* *in der ganzen Welt*
all right *42* *verstanden*
all the time *32* *die ganze Zeit*
all together *69* *insgesamt*
Allen key *16* *Inbusschlüssel, Innensechskantschlüssel*
allergic *101* *allergisch*
allow *63* *rechnen mit*
along *36* *entlang*
alphabetical *94* *alphabetisch*
alternative *111* *Alternative*
aluminium *83* *Aluminium*
amazing *136* *erstaunlich*
ambient *99* *Umgebungs-*
ambulance *61* *Krankenwagen*
among other things *127* *unter anderem*
amount *77* *Menge*
amusing *82* *amüsant*
and so on *135* *und so weiter*
angle *92* *Winkel*
angle grinder *16* *Winkelschleifer*
angry *43* *böse, wütend, zornig*
annoyed *64* *verärgert*
annual *47* *jährlich*
answer *73* *beantworten*
answerphone *10* *Anrufbeantworter*
antenna *137* *Antenne*
anticlockwise *98* *gegen den Uhrzeigersinn*
anti-freeze *64* *Frostschutz(mittel)*
anti-shoplifting *136* *gegen Ladendiebstahl*
any more *42* *noch irgendwelche*
any other business *90* *Verschiedenes*
anyhow *42* *jedenfalls*
apart *94* *getrennt*
apex *92* *Spitze, Scheitel*
appliance *78* *Gerät*
applicant *42* *Bewerber/in*
application *48* *Bewerbung*
applied science *49* *angewandte Wissenschaft*
apply *48* *bewerben*
appoint *127* *ernennen*
appointment *68* *Termin, Verabredung*
apprentice *6* *Lehrling, Auszubildende/r*
apprenticeship *42* *Lehre, Ausbildung(splatz)*
appropriate *62* *entsprechend*
approx *31* *ungefähr, zirka*
area *39* *Bereich*
argue *54* *(sich) streiten*
argument *110* *Argument, Einwand*
around the world *124* *auf der ganzen Welt*
arrange *73* *anordnen*
arrangement *63* *Termin, Vereinbarung*
array *113* *Ansammlung, Reihe*
arrival *128* *Ankunft*
arrowhead *98* *Pfeilspitze*
as a rule *26* *in der Regel*
as a whole *115* *insgesamt*
as far as ... goes *119* *was ... anbetrifft*
asleep, be ~ *81* *schlafen*
as well as *78* *sowie*
as well *29* *auch, ebenfalls*
ask the way *36* *nach dem Weg fragen*
assembly *29* *Montage, Zusammenbau*
assembly line *83* *Montage-, Fließband*
assembly plant *42* *Montage-, Herstellwerk*
assignment *98* *Aufgabe*
assist *146* *helfen, unterstützen*
assistant *33* *Assistent/in*
associate *47* *Partner/in*
at a time *60* *auf einmal*
at all *42* *überhaupt*
at fault *28* *schuldig*
at first *60* *zuerst, anfangs*
at last *60* *schließlich*
at least *33* *mindestens*
at once *30* *sofort*

at present *48* *momentan, gegenwärtig*
at the moment *6* *im Augenblick*
atmosphere *99* *Atmosphäre*
attach *48* *beilegen, -fügen*
attack *129* *angreifen*
attendant *38* *Wächter/in*
attract *69* *anlocken, gewinnen*
attraction *47* *Anreiz*
audience-friendly *127* *publikumsfreundlich*
Austria *6* *Österreich*
Austrian *6* *Österreicher/in; österreichisch*
authentic *116* *authentisch*
authorised dealer *100* *Vertragshändler*
automatic *25* *automatisch*
automatically *129* *automatisch*
automotive *10* *Auto(mobil)-*
auxiliary *99* *zusätzlich*
available *62* *verfügbar*
average *49* *durchschnittlich*
aviation *122* *Luftfahrt, Flug-*
avoid *31* *ausweichen*
aware *119* *bewusst, aufmerksam*
awareness *123* *Bewusstsein*
awful *36* *furchtbar*
axis *121* *Achse*

B

back to back *89* *Rücken an Rücken*
backup *132* *Sicherungskopie*
bad luck *30* *Unglück, Pech*
balcony *103* *Balkon*
ban *122* *verbieten, verzichten auf*
bank *22* *Ufer*
bank robbery *130* *Bankraub*
bar chart *120* *Säulen-, Balkendiagramm*
bar code *9* *Strichkodierung*
barge *122* *Last-, Binnenschiff*
base *87* *Grund-*
basic *11* *grundlegend, elementar*
basically *121* *grundsätzlich*
batch *142* *Gruppe*
bathroom *66* *Bad(ezimmer)*
battery *16* *Batterie*
bay *132* *Ausgang, Anschluss*
beam *85* *Balken*
beautiful *127* *schön*
bedsitter *66* *möbliertes Zimmer*
behaviour *120* *Verhalten, Benehmen*
belong *58* *gehören*
belt sander *16* *Schwingschleifer*
bend *84* *(sich) biegen (lassen)*
beneath *80* *unter(halb)*
benefit *100* *Vorteil, Nutzen*
bent *81* *gekrümmt, angewinkelt*
between *22* *zwischen*
bicycle *106* *Fahrrad*
bike *74* *Fahrrad*
billion *120* *Milliarde*
binary *126* *binär*
biometric ID card *130* *biometrischer Ausweis*
birth *49* *Geburt*
block *65* *blockieren*
blunt *20* *stumpf*
boast *38* *prahlen, angeben*
boiling point *127* *Siedepunkt*
bolt *77* *Bolzen, Schraube*
bond *140* *verbinden, verkleben*
bonding *85* *Verklebung*
boom *71* *Hochkonjunktur*
border *59* *Grenze*
bored *53* *gelangweilt*
boring *82* *langweilig*
borough *139* *Stadtbezirk*
borrow *74* *(sich) (aus)leihen*
boss *43* *Chef/in*
bottle *83* *Flasche*
bottom *34* *(unteres) Ende, unten*
bottom-hung window *98* *Kippfenster*
bottom out *121* *die Talsohle erreichen*
boundary *124* *Grenze*
boutique *131* *Boutique*
box *6* *Kasten, Kästchen*
brainstorm *83* *(Ideen) sammeln (zu)*
brainstorming *140* *Ideensammeln*
brakes *64* *Bremsen*
brass *84* *Messing*
Brazil *127* *Brasilien*
break *59* *Pause;* *84* *(zer)brechen*
breakdown *12* *Panne*
breakthrough *143* *Durchbruch*
bright *142* *hell, strahlend*
Britain *6* *Großbritannien*
British *6* *Brite/Britin; britisch*
brittle *84* *brüchig, spröde*
broken *12* *kaputt, defekt*
brown *81* *braun*
buckled *25* *verbogen*
build *96* *bauen, errichten*
building *39* *Gebäude, Bau*
bundle *142* *Bündel, Paket*
burn *99* *verbrennen*
burning *120* *Verbrennung, Brand*
bus station *34* *Busbahnhof*
bus stop *18* *Bushaltestelle*
business people *139* *Geschäftsleute*
business premises *133* *Geschäftsräume, Firmengelände*
business travel *88* *Geschäftsreisen*
busy *60* *besetzt, belebt*
butt *146* *Fass, Tonne*
button *44* *Taste, Knopf*
buyer *133* *Käufer/in*
by far *118* *bei weitem*
by law *78* *gesetzlich*

C

cable *44* *Kabel*
cabling *131* *Verkabelung*
calculate *60* *berechnen*
calculation *11* *Rechenart, -operation*
call *66* *anrufen*
call a meeting *91* *eine Sitzung anberaumen*
call up *78* *aufrufen*
caller *115* *Anrufer/in*
calm *78* *ruhig*
camera *38* *Fotoapparat, Kamera*
campsite *97* *Campingplatz*
cancel *96* *absagen*
candle *142* *Kerze*
canteen *21* *Kantine, Cafeteria*
capacity *117* *Kapazität, Fassungsvermögen*
caption *40* *Bildunterschrift*
car body *21* *Karosserie*
car crash *21* *Autounfall*
car ferry *58* *Autofähre*
car park *37* *Parkplatz, Parkhaus*
car wash *134* *Autowaschanlage*
carbon *115* *Kohlenstoff, -dioxid*
carbon-intensive *124* *kohlenstoffintensiv*
cardboard *40* *Pappe*
care *71* *betreuen*
career chances *46* *Berufschancen*
careers advice *46* *Berufsberatung*
careful *27* *sorgfältig*
carpenter *76* *Zimmermann, Tischler/in*
carry out *64* *durchführen*
carton *44* *Karton*
case *30* *Fall*
cash desk *60* *Kasse*
cashier *134* *Kassierer/in*
casing *25* *Gehäuse*
catalogue *148* *Katalog*
catalyst *107* *Katalysator*
catch *28* *(auf)fangen*
category *47* *Kategorie, Fachgebiet*
cause *28* *verursachen*
cautious *82* *vorsichtig*
CCTV camera *130* *Überwachungskamera*
CD player *38* *CD-Spieler*
ceiling *85* *(Zimmer-)Decke*
cellar *103* *Keller*
cellphone *138* *Handy, Mobiltelefon*
central heating programmer *144* *Zentralheizungsregler*
centre of attention *82* *Mittelpunkt der Aufmerksamkeit*
centre punch *16* *Körner*
ceramic *83* *Keramik*
certain *96* *sicher*
certainly *82* *sicherlich*
certificate *49* *Zeugnis*
CFC *101* *FCKW*
chain of events *65* *Kette von Ereignissen*
chair *85* *Stuhl;* *144* *leiten*
change *48* *(sich) verändern;* *48* *(Ver-)Änderung*
channel *132* *Kanal*

Channel tunnel *112* *Kanaltunnel*
character *25* *Buchstabe*
characteristic *99* *Eigenschaft*
charger *114* *Aufladevorrichtung*
charming *82* *charmant*
chat *60* *sich unterhalten*
check *25* *(über)prüfen, kontrollieren*
checklist *89* *Checkliste*
cheerio *70* *tschüs, servus*
chemical *77* *Chemikalie*
chemical fixing agent *85* *chemischer Klebstoff*
chemical reaction *109* *chemische Reaktion*
chemist *6* *Chemiker/in*
chill out *54* *sich entspannen, abhängen*
chin *80* *Kinn*
chisel *16* *Meißel, Stemmeisen*
chlorofluorocarbon *101* *Fluorchlorkohlenwasserstoff*
choice *52* *Wahl, Auswahl*
chores *72* *Routinearbeit, Arbeiten im Haushalt*
Christmas tree *142* *Weihnachtsbaum*
chuck *98* *Bohrfutter*
chuckle *80* *in sich hineinlachen, kichern*
church *36* *Kirche*
circle *82* *Kreis*
circuit *49* *(Strom-)Kreis*
circulation *101* *Zirkulation*
cistern *98* *Spülkasten*
civil rights *135* *Bürgerrechte*
civil *127* *zivil*
clasped *80* *gefaltet*
class *49* *Klasse*
clean *33* *reinigen, säubern*
clear *50* *klar, deutlich*
clever *53* *klug*
click *78* *(an)klicken*
climate *118* *Klima*
climate change *79* *Klimawandel*
climb *33* *klettern, steigen;* *121* *klettern, ansteigen*
clip *85* *Klammer, Lasche, Schelle*
clock on *23* *einstempeln*
clock-face *94* *Zifferblatt*
clockwise *98* *im Uhrzeigersinn*
close *28* *nahe*
close *38* *(ab)schließen*
close down *128* *schließen, abschalten*
close to home *124* *nahe am Haus*
closed-circuit television *130* *Überwachungskamera(s)*
clothes *145* *Kleidung*
clothing *136* *Bekleidung, Kleider*
cloud *97* *Wolke*
clue *50* *Hinweis*
clutch *21* *Kupplung*
CNC lathe operator *76* *Dreher/in an einer computergesteuerten Drehmaschine*
coal *79* *Kohle*
coat *142* *beschichten*
coffee maker *23* *Kaffeemaschine*
coil *99* *Schlange*
collar *98* *Ring*
collect *43* *(ab)holen*
Cologne *6* *Köln*
colour *81* *Farbe*
colourful *82* *bunt, farbig*
column drill *17* *Ständerbohrmaschine*
combination *92* *Kombination, Verbindung*
combination boiler *98* *Wasserkocher*
combine *92* *zusammenfügen, kombinieren*
come on *60* *beeil dich*
comfort *101* *Komfort*
comfortable *58* *bequem*
comment *53* *Bemerkung*
commercial lawyer *135* *Wirtschaftsanwalt*
common *10* *normal, üblich*
communication *135* *Kommunikation*
communications *78* *Nachrichtenübermittlung*
community *118* *Gemeinde*
compact *66* *kompakt, klein*
companionship *72* *Gesellschaft*
company *23* *Unternehmen, Firma*
compare *83* *vergleichen*
comparison *117* *Vergleich, Steigerung*
compete *107* *konkurrieren*
competitive *42* *konkurrenzfähig*
compile *119* *zusammenstellen*
complaint *89* *Beanstandung, Reklamation*
complete *17* *vervollständigen*
completely *75* *völlig*
complicated *25* *kompliziert*
component *40* *(Einzel-)Teil*
composite *83* *zusammengesetzt*
compound *21* *Zusammensetzung*
comprehensive school *49* *Gesamtschule*
compressed *99* *komprimiert*
compression *132* *Kompression*
compressor *99* *Kompressor*
comprise *132* *umfassen*
computer-aided design *43* *computergestützte Konstruktion*
concentrate on *71* *sich konzentrieren auf*
conclusion *72* *(Schluss-)Folgerung, Entscheidung*
concrete *29* *Beton*
condensing boiler *144* *Kondenskessel*
condensor *99* *Kondensator*
condition *65* *Zustand*
conduction element *102* *Leitungselement*
conductor *141* *Leiter*
cone *28* *Warnhütchen*
conference room *50* *Besprechungszimmer, Konferenzraum*
connect *44* *anschließen*
connection *78* *Verbindung, Anschluss*
connector *75* *Bindewort*
cons *72* *Wider, Kontra*
consequence *120* *Folge, Konsequenz*
conservation *127* *Erhaltung*
conserve *146* *erhalten*
consider *72* *Rücksicht nehmen auf*
considerably *101* *beträchtlich*
consignment *104* *(Waren-)Sendung, Lieferung*
consist of *113* *bestehen aus*
constant *101* *gleich bleibend*
construction *6* *Bau*
construction site *105* *Baustelle*
consume *99* *verbrauchen*
consumer *138* *Verbraucher/in*
consumption *114* *Verbrauch*
contact *6* *Kontakt, Verbindung*
contain *142* *enthalten*
contaminated *109* *verunreinigt*
contents *44* *Inhalt*
continent *58* *(europäischer) Kontinent*
continental plate *124* *Kontinentalplatte*
continue *62* *fortsetzen*
continuous *131* *ständig*
contract *96* *Vertrag, Auftrag*
contractor *52* *Handwerksbetrieb, Bauunternehmen*
contribution *48* *Beitrag*
controller *113* *Steuergerät*
controversy *111* *Kontroverse*
convenience *100* *Komfort, Bequemlichkeit*
convenient *135* *bequem*
conventional *101* *herkömmlich, konventionell*
conversion *107* *Umwandlung*
convert *66* *umbauen*
conveyor *33* *Aufzug*
cook *43* *kochen*
cooker *78* *Herd, Kocher*
cool *99* *kühl*
cooling fan *141* *Kühllüfter*
copier *27* *Kopierer*
core *124* *Kern*
corner *28* *Ecke*
correct *12* *richtig*
correct *120* *korrigieren*
cost of living *72* *Lebenshaltungskosten*
cottage *66* *kleines (Land-)Haus, Ferienhaus*
cotton *83* *Baumwolle*
countable *86* *zählbar*
counterclockwise *98* *gegen den Uhrzeigersinn, links herum*

countryside *115* *Landschaft*
couple *66* *Ehepaar*
course *6* *Kurs, Lehrgang*
cover *101* *bedecken, einschließen*
covering *102* *Belag, Verkleidung*
covers *81* *Bettdecke*
crack *84* *knacken, brechen, springen*
craft *62* *Fahrzeug, Schiff*
crazy *141* *verrückt*
credit card *135* *Kreditkarte*
credit crunch *116* *Kreditknappheit, -klemme*
crime *48* *Kriminalität*
criminal lawyer *135* *Anwalt für Strafsachen*
crisis *127* *Krise*
critical *25* *kritisch*
criticise *111* *kritisieren*
Croatia *6* *Kroatien*
Croatian *6* *Kroate/Kroatin; kroatisch*
cross *59* *überqueren*
crowd *87* *(Menschen-)Menge*
cruel *130* *grausam*
crust *124* *Kruste*
crystal *142* *Kristall*
cupboard *149* *Schrank*
curled *80* *untergeschlagen*
current *90* *derzeitig, aktuell*
curved *92* *gekrümmt*
curvilinear *92* *gekrümmt*
customer *23* *Kunde/Kundin*
Customer Services *90* *Kundendienst*
cut out *25* *(sich) ausschalten*
CV *49* *Lebenslauf*
cycle track *122* *Fahrradweg*
cycling *27* *Radfahren*
cylinder *21* *Zylinder*
cylindrical *77* *zylindrisch*
Czech *6* *Tscheche, Tschechin; tschechisch*
Czech Republic *6* *Tschechische Republik, Tschechien*

D

damage *25* *beschädigen*
dancing *54* *Tanzen*
danger *108* *Gefahr*
dangerous *78* *gefährlich*
dashboard *61* *Armaturenbrett*
data processing *41* *Datenverarbeitung, EDV*
databank *138* *Datenbank*
database *130* *Datenbank*
day after day *23* *tagaus, tagein*
day before yesterday *32* *vorgestern*
daylight *142* *Tageslicht*
daytime *49* *tagsüber*
dazzle *65* *blenden*
DC *109* *Gleichstrom*
deactivate *136* *deaktivieren*
deal with *121* *sich befassen mit, behandeln*
dealer *44* *Händler/in*
death *142* *Tod*
debate *120* *Debatte*
decay *124* *Zerfall*
decide *42* *entscheiden*
decision *82* *Entscheidung*
decrease *121* *Rückgang, Abnahme*
deep *124* *tief*
definitely *67* *bestimmt, sicherlich*
definition *72* *Definition*
deforestation *120* *Abholzung, Entwaldung*
degree *92* *Grad*
delay *88* *Verspätung, Verzögerung*
deliver *41* *anliefern*
delivery *34* *Zustellung, Lieferung*
demand *78* *Nachfrage, Bedarf*
demanding *53* *anspruchsvoll*
dental appointment *148* *Zahnarzttermin*
department *33* *Abteilung*
department store *131* *Kaufhaus, Warenhaus*
depend on *25* *abhängen von*
dependence *146* *Abhängigkeit*
dependency on *111* *Abhängigkeit von*
deposit *66* *Anzahlung, Kaution*
depth *94* *Tiefe*
describe *22* *beschreiben*
description *25* *Beschreibung*
design *77* *entwerfen, entwickeln*
destroy *115* *zerstören, vernichten*
destructive *118* *zerstörerisch*
detailed *103* *detailliert, ausführlich*
deteriorate *107* *(sich) verschlechtern*
determine *126* *bestimmen*
devastate *118* *zerstören, verwüsten*
develop *108* *entwickeln*
developed world *138* *Industrieländer*
device *47* *Vorrichtung, Gerät*
dialogue *8* *Dialog*
diameter *124* *Durchmesser*
diary *64* *Terminkalender, Tagebuch*
dictate *9* *diktieren*
dictionary *135* *Wörterbuch*
differ *126* *sich unterscheiden*
difference *88* *Unterschied*
dig *124* *graben*
digital measuring slide *16* *digitaler Messschieber*
dimension *92* *Abmessung, Maß*
direct current *107* *Gleichstrom*
direct *44* *direkt*
direction *61* *Richtung*
directions *34* *Wegbeschreibung*
dirty *128* *schmutzig*
disadvantage *72* *Nachteil*
disagree *122* *anderer Meinung sein, widersprechen, nicht zustimmen*
disappoint *42* *enttäuschen*
disaster *110* *Katastrophe, Unglück*
discount *79* *Rabatt, Skonto, Nachlass*
discovery *142* *Entdeckung*
discreet *131* *diskret*
discuss *52* *besprechen, diskutieren (über)*
discussion *90* *Diskussion, Erörterung*
dishwasher *98* *Geschirrspüler*
dislike *139* *nicht mögen*
dispatch *40* *Versand, Auslieferung*
disposal *115* *Entsorgung*
dissipate *141* *ableiten*
distance *19* *Abstand*
distribute *79* *(aus)liefern, verteilen*
distribution *78* *Verteilung, Versorgung*
divide into *86* *unterteilen in*
do well *48* *gut abschneiden, gut gehen*
docks *23* *Hafen(anlagen)*
document *38* *Dokument, Papier*
documentation *89* *Dokumentation, Unterlagen*
dominant *82* *beherrschend, dominant*
don't mention it *36* *nichts zu danken*
Dos and Don'ts *38* *Gebote und Verbote*
double *110* *verdoppeln*
double doors *43* *Doppeltür*
doubt *115* *bezweifeln*
download *132* *herunterladen*
downtime *129* *Ausfallzeit*
draughtsperson *6* *Zeichner/in*
draw up *89* *aufstellen*
drawer *18* *Schub(lade)*
drawing *40* *Zeichnung*
dream *81* *Traum; träumen*
dreamless *81* *traumlos*
drill *16* *Bohrer, Bohrmaschine;* *125* *bohren*
drill bit *86* *Bohrer*
drive in *19* *einschrauben, eindrehen*
driving licence *49* *Führerschein*
drop *88* *Rückgang;* *99* *fallen*
drought *118* *Trockenzeit, Dürre(periode)*
drown *118* *ertrinken, ertränken*
drug *65* *Droge, Rauschgift*
dry *101* *trocken*
duct *99* *Rohr*
ductile *84* *(ver)formbar, ausziehbar*
due to *101* *wegen, aufgrund von*
durable *84* *haltbar, widerstandsfähig, langlebig*
during *40* *während*
dust *101* *Staub*
dust mite *101* *(Haus-)Staubmilbe*
Dutch *6* *Holländer/in; holländisch*
duty *50* *Aufgabe, Pflicht*
dynamo *87* *Lichtmaschine, Dynamo*

E

each other *6* *einander, sich*
ear *80* *Ohr*
ear muffs *149* *Ohrenschützer*
earn *129* *verdienen*
earth *124* *Erde*
earthquake *124* *Erdbeben*
ease *100* *Leichtigkeit*
Easter *32* *Ostern*
eastern *59* *Ost-*
echo *7* *Echo*
ecological *73* *ökologisch*
ecologically neutral *101* *ökologisch neutral*
ecologically friendly *141* *umweltfreundlich*
economical *129* *sparsam, wirtschaftlich*
economy *62* *Touristenklasse*
edge *124* *Rand*
education *49* *(Schul-)Bildung, Ausbildung*
effect *103* *(Aus-)Wirkung, Einfluss*
effective *100* *effektiv, wirkungsvoll*
efficient *43* *leistungsfähig*
either way *119* *in jeder/keiner Hinsicht*
elastic *102* *elastisch*
electric charge *142* *elektrische Ladung*
electrical *25* *elektrisch, Strom-*
electrical charge *109* *elektrische Ladung*
electrically-operated *78* *mit Elektroantrieb, elektrisch betrieben*
electrician *6* *Elektriker/in, Elektrotechniker/in*
electricity *78* *Elektrizität, Strom*
electrochemical *108* *elektrochemisch*
electrolyte *108* *Elektrolyt*
electron *109* *Elektron*
electronic article surveillance *136* *elektronische Artikelüberwachung*
electronics *6* *Elektronik*
ellipse *94* *Ellipse*
emerge *126* *austreten*
emergency *105* *Notfall*
emissions *108* *Abgase*
emit *120* *emittieren, ausstoßen*
emphasis *146* *Betonung*
emphasise *100* *betonen, hervorheben*
employ *47* *beschäftigen*
employer *80* *Arbeitgeber/in, Unternehmer/in*
employment agency *46* *(private) Stellenvermittlung*
employment prospects *78* *Aussichten auf Arbeit, Beschäftigungschancen*
enable *113* *(es) ermöglichen*
encyclopedia *135* *Enzyklopädie, Lexikon*
energy *108* *Energie*
engine *55* *Motor*
engineer *6* *Ingenieur/in, Techniker/in*
engineering *22* *Technik, Maschinenbau*
English-speaking *55* *englischsprachig*
enhanced *132* *verbessert*
enjoy *8* *gern haben/tun*
enquiry *132* *Anfrage*
ensure *78* *sicherstellen, gewährleisten*
enter *109* *eindringen (in)*
entrance *68* *Eingang*
environment *29* *Umgebung, Umfeld*
environmental *100* *Umwelt-*
environmental pollution *120* *Umweltverschmutzung*
environmentally friendly *100* *umweltfreundlich*
equal *92* *gleich*
equilateral triangle *93* *gleichseitiges Dreieck*
equip *79* *ausrüsten, ausstatten*
equipment *23* *Ausrüstung, Geräte*
equivalent *24* *Entsprechung*
erection *78* *Errichtung*
erupt *124* *ausbrechen*
escape *122* *entweichen*
essential *101* *(absolut) notwendig*
estate *73* *Kombi*
estimate *143* *schätzen*
European *47* *europäisch*
evaluate *90* *aus-, bewerten*
evaporator coil *99* *Verdampferschlange*
even if *128* *selbst wenn*
evening class *129* *Abendschule*
ever since *139* *seitdem*
everyday life *54* *Alltag*
evidence *87* *Beweismaterial, -mittel*
evidently *104* *offensichtlich*
exact *19* *genau*
exaggerated *119* *übertrieben*
exam *52* *Prüfung, Examen*
examine *82* *untersuchen, prüfen*
except *42* *außer*
exceptionally *99* *außerordentlich*
excess *109* *überschüssig, überzählig*
excited *142* *erregt*
exciting *82* *aufregend*
exclusive *131* *exklusiv*
exercise book *84* *(Schul-)Heft*
exhaust outlet *40* *Auspuff(rohr)*
exist *125* *existieren*
exit *60* *Ausfahrt, Ausgang*
expand *47* *sich ausdehnen, expandieren*
expect *59* *erwarten*
expectation *48* *Erwartung*
expenses *148* *Spesen(abrechnung)*
experience *46* *(Berufs-)Erfahrung*
expert *91* *Fachmann/frau, Experte/Expertin*
expert opinion *118* *Expertenmeinung*
explain *100* *erklären, erläutern*
explanation *98* *Erklärung*
explore *140* *erforschen, erkunden*
explosion *108* *Explosion*
export *47* *Ausfuhr, Export*
expose to *143* *aussetzen*
express *10* *ausdrücken*
expression *24* *Ausdruck*
extract *31* *Auszug;* ***99*** *gewinnen, entziehen*
extractor fan *33* *Entlüfter*
extreme *81* *Extrem*
extremely *82* *äußerst, außerordentlich*
eyesight *78* *Sehvermögen*

F

face *139* *Gesicht*
factory *6* *Fabrik*
factual *31* *sachlich*
fail *139* *scheitern, (es) nicht schaffen*
fall into *92* *sich unterteilen lassen*
famous *142* *berühmt*
far too *65* *bei weitem*
fare *62* *Fahrpreis*
farmland *118* *landwirtschaftlich genutzte Fläche*
fashion *131* *Mode*
fast food *96* *Schnellgerichte*
fast lane *65* *Überholspur*
fault *125* *(Verwerfungs-)Spalte*
faulty *25* *fehlerhaft, defekt*
favourable *48* *positiv, günstig*
favourite *129* *Lieblings-*
fax machine *105* *Faxgerät*
feature *100* *Merkmal*
federal *123* *Bundes-*
feed into *108* *einspeisen*
feeling *135* *Gefühl*
ferrous *115* *Eisen-*
ferry *22* *Fähre*
ferry port *59* *Fährhafen*
fetch *33* *holen*
fibreglass *83* *Glaswolle, -faser*
fight *81* *kämpfen*
figure *31* *Zahl, Ziffer*
filament *141* *(Glüh-)Faden*
file *16* *Feile*
fill in *12* *eintragen, -setzen*
filling station *134* *Tankstelle*
filter *33* *Filter;* ***60*** *einspurige Auffahrt, Abbiegespur;* ***146*** *filtern*
finalise *89* *(endgültig) festlegen*
find out *23* *feststellen*
findings *127* *Ergebnisse*
fire engine *112* *Feuerwehrauto*
firm *14* *Firma*
first floor *103* *erste Etage, erster Stock*
first of all *52* *zu(aller)erst, zunächst*

first-aid kit *97* *Verbandskasten*
firstly *110* *zuerst*
fit *40* *passen*
fitted carpet *103* *Teppichboden*
fix *28* *befestigen;* ***145*** *Lösung*
flash *31* *blinken*
flash steam plant *126* *Heißdampfkraftwerk*
flick over *61* *weiterrücken*
flight *148* *Flug*
float *81* *schweben*
flood *118* *Flut, Überschwemmung*
flood defences *120* *Dämme*
floor *29* *(Fuß-)Boden*
flooring *104* *Bodenbelag*
flow *113* *fließen*
fluctuation *119* *Schwankung*
fluid *115* *Flüssigkeit*
fluorescent *141* *Leucht-, fluoreszierend*
flush *146* *spülen*
fly *81* *fliegen*
foam rubber *83* *Schaumgummi*
focus *83* *Brennpunkt, Schwerpunkt*
foil *101* *Folie*
folded *80* *verschränkt*
fool *61* *Dummkopf*
foot contact *141* *Fußkontakt*
football club *49* *Fußballverein*
football ground *34* *Fußballplatz*
football match *45* *Fußballspiel*
for example *53* *zum Beispiel*
for God's sake *29* *um Gottes willen*
force *44* *Gewalt;* ***110*** *zwingen*
forecast *97* *(Wetter-)Vorhersage*
forest *118* *Wald*
forklift (truck) *29* *Gabelstapler*
formal *133* *formal, förmlich*
fortune *71* *Vermögen*
forum *138* *Forum*
fossil fuel *120* *fossiler Brennstoff*
fractured *125* *brüchig*
frame *147* *Rahmen*
frames per second (FPS) *132* *Bilder pro Sekunde*
France *23* *Frankreich*
freezer *134* *Gefriertruhe, Tiefkühltruhe*
French *6* *Franzose/Französin; französisch*
frequency *120* *Häufigkeit*
frequent *108* *häufig, oft*
fresh *82* *frisch*
friendly *72* *freundlich, freundschaftlich*
frighten away *131* *abschrecken, verscheuchen*
from the start *71* *von Anfang an*
fuel cell *107* *Brennstoffzelle*
fuel mix *21* *Kraftstoffgemisch*
fuel tank *55* *Treibstofftank*
fully *48* *völlig*
fully-wired *146* *voll verkabelt*
fumarole *124* *vulkanische Dampfquelle*
function *19* *Funktion;* ***99*** *funktionieren, arbeiten*
funny *80* *lustig, witzig*
furnished *66* *möbliert, eingerichtet*
furniture *6* *Möbel*
furthermore *75* *ferner, zudem*
future-orientated *77* *zukunftsorientiert*

G

gadget *146* *Vorrichtung, Gerät*
game *10* *Spiel*
gap *38* *Lücke*
garage *66* *Garage*
gaseous *99* *gasförmig*
gas-fired *99* *Gasheiz(kessel)*
gasoline *123* *Benzin*
gate *35* *Tor*
general *52* *allgemein*
generally *26* *im Allgemeinen*
generate *79* *erzeugen*
generation *49* *Erzeugung*
generator *78* *Generator*
generous *47* *großzügig*
geologist *125* *Geologe/Geologin*
geothermal *124* *Erdwärme-*
German *6* *Deutsche/r; deutsch*
Germany *6* *Deutschland*
gesture *100* *Geste*
get across *100* *vermitteln*
get dressed *23* *sich anziehen*
get from A to B *116* *von A nach B gelangen*
get on with *30* *erledigen*
get results *43* *Erfolg haben*
get rid of *115* *abschaffen*
get to know *6* *kennen lernen*
get up *23* *aufstehen*
getaway car *130* *Fluchtwagen*
geyser *124* *Geysir*
girlfriend *27* *Freundin*
give rise to *142* *veranlassen, erregen*
glacier *118* *Gletscher*
glad *135* *froh*
glass *83* *Glas*
global *42* *global, weltweit*
global positioning system *130* *GPS*
global village *46* *globales Dorf*
global warming *118* *Erderwärmung, Treibhauseffekt*
glow *142* *Leuchten, Schein*
glue *85* *Klebstoff, Leim*
go ahead *6* *fortfahren*
go clubbing *54* *in Diskos/Nachtklubs gehen*
go on *29* *fortfahren*
go out in the field *53* *ins Freie gehen*
go to sleep *81* *einschlafen*
go well *116* *gut (aus)gehen*
gold *83* *Gold*
good heavens *137* *ach du meine Güte*
good luck *36* *viel Glück*
goods inward *40* *Warenannahme*
goods outward *41* *Warenauslieferung*
GPS tracking system *130* *GPS Satellitennavigationssystem*
grab handle *73* *Haltegriff*
grade *49* *Note*
gradually *121* *allmählich, schritt-, stufenweise*
graduate *142* *Hochschulabsolvent/in*
Greece *6* *Griechenland*
Greek *6* *Grieche, Grieching; griechisch*
green *36* *grün*
greeting *90* *Begrüßung*
grey *81* *grau*
greywater *146* *Brauchwasser*
grid *31* *Tabelle*
group *6* *Gruppe*
grow *53* *wachsen*
growth *47* *Wachstum, Zunahme*
guarantee *29* *garantieren*
guard rail *28* *Schutzgeländer*
guess *88* *(er)raten*
guest *145* *Gast*
guide *8* *Anleitung;* ***101*** *leiten*
guy *29* *Typ, Junge*

H

hacksaw *16* *Bügelsäge*
hair *80* *Haar/e*
half past seven *60* *halb acht*
halfway *113* *halb*
halogen bulb *141* *Halogenlampe*
hammer *12* *Hammer*
handball coach *82* *Handballtrainer/in*
handheld *136* *Hand-(Gerät)*
handle *137* *umgehen mit, erfassen*
handling *116* *Handhabung*
handset *44* *Handapparat*
hang around *54* *herumhängen*
happen *29* *passieren, geschehen*
harbour *69* *Hafen*
hard copy *23* *Ausdruck*
hard disk *77* *Festplatte*
hard hat *28* *Schutzhelm*
hard shoulder *61* *Seitenstreifen, Standspur*
harmful *108* *schädlich*
harness *126* *nutzbar machen*
have confidence in *82* *Vertrauen haben zu*
have sth against *136* *etw dagegen haben*
have to do with *21* *zu tun haben mit*
head office *47* *Zentrale*
headline *140* *Schlagzeile, Überschrift*
headquarters *46* *Firmensitz, Zentrale*

irregular *93* *unregelmäßig*
isosceles *92* *gleichschenklig*
it is no surprise that *108* *es ist kein Wunder, dass*
Italian *6* *Italiener/in; italienisch*
Italy *6* *Italien*
item *9* *Punkt, Gegenstand, Eintrag*

J

jaw *98* *Bohrfutter*
jeweller *131* *Juwelier/in*
jewellery *149* *Schmuck*
jig saw *16* *Stichsäge*
job centre *46* *Arbeitsagentur*
job description *78* *Stellenbeschreibung*
job offer *46* *Stellenangebot*
join *19* *verbinden*
joint *77* *Naht(stelle)*
journey *23* *Fahrt*
judgement *82* *Urteil*
junior certificate *49* *Mittlere Reife*
just about *143* *fast*
just for fun *33* *nur zum Spaß*
just the thing *58* *genau das Richtige*

K

Kazach *6* *Kasache/Kasachin; kasachisch*
Kazachstan *6* *Kasachstan*
keep away from *149* *sich fernhalten von*
keep to *42* *sich halten an*
key point *100* *Kernpunkt*
key word *73* *Schlüsselwort*
keyed *98* *mit Schlüssel*
keyless *98* *ohne Schlüssel*
kid oneself *120* *sich etw vormachen*
kinetic *110* *kinetisch*
kiss *30* *leicht berühren*
knee *80* *Knie*
knife *20* *Messer*
knock in *19* *(hin)einschlagen*

L

lab assistant *6* *Laborassistent/in*
label *136* *Schild(chen), Etikett*
labour-saving *146* *arbeitssparend*
lamp-holder *85* *Lampenfassung*
land *59* *landen*
lane *36* *Gasse, Weg*
largely *119* *zum größten Teil, weit gehend*
laser beam *142* *Laserstrahl*
latest *91* *neueste, letzte*
lathe operator *6* *Dreher/in*
laugh *80* *lachen*
lava *124* *Lava*
lay *85* *verlegen*
lay off *87* *(zeitweilig) entlassen*
layer *102* *Schicht*
lead *110* *Blei*
lead to *88* *führen zu*
leaflet *38* *Broschüre, Prospekt*
leather *83* *Leder*
legendary *110* *legendär*
leisure centre *37* *Freizeitzentrum*
length *19* *Länge*
let's see *51* *mal sehen*
letter to the editor *135* *Leserbrief*
letting agency *71* *(Immobilien-) Makler/in*
lever *129* *Hebel*
lie *81* *liegen*
lifestyle *96* *Lebensstil*
lift *96* *Aufzug, Fahrstuhl*
light bulb *33* *Glühbirne*
light fitting *131* *Lampenfassung*
lighting *25* *Beleuchtung*
like the look of *133* *gefallen das Aussehen*
limited *10* *mit beschränkter Haftung*
limitless *109* *unbegrenzt, grenzenlos*
line graph *120* *(Kurven-, Linien-) Diagramm*
link *7* *verbinden*
liquid *99* *flüssig; Flüssigkeit*
liquify *99* *flüsssig werden, (sich) verflüssigen*
list price *133* *Listenpreis*
listener *140* *Zuhörer/in*
litre *116* *Liter*
lively *82* *lebendig*
living area *66* *Wohnfläche*
living room *18* *Wohnzimmer*
load *109* *Ladung*
loaded *101* *beladen*
loading ramp *41* *Laderampe*
lobby *131* *Foyer, Eingangshalle*
local *71* *örtlich, lokal*
locate *31* *platzieren*
location *22* *Lage*
lock *30* *feststellen, arretieren*
lock away *136* *wegschließen*
locking castor *28* *feststellbare Laufrolle*
locking jack *31* *Karabinerhaken*
logical *40* *logisch*
logistics *25* *Logistik, Auslieferung*
lonely *38* *einsam*
look at *17* *ansehen, betrachten*
look for *55* *suchen nach*
look like *23* *aussehen wie*
loop *132* *Schleife*
loose *149* *weit (geschnitten), locker*
loosen *77* *lockern, lösen*
lose *64* *verlieren*
lose weight *96* *abnehmen*
loser *119* *Verlierer/in*
loss *101* *Verlust*
lottery *9* *Lotterie, Lotto*
low-energy light bulb *111* *Sparlampe*
low-lying *118* *niedrig liegend*
loyal *82* *treu, loyal*
loyalty *82* *Treue, Loyalität*
Ltd *132* *(Gesellschaft) mit beschränkter Haftung, GmbH*
luckily *34* *glücklicherweise*
luggage *59* *Gepäck*
lumpy *142* *klumpig*
luxury *107* *Luxus*

M

machine *23* *Maschine*
machine operator *6* *Maschinenarbeiter/in*
machinery *52* *Maschinen, Geräte*
made in *27* *hergestellt in*
made up of *42* *bestehend aus*
magazine *134* *Zeitschrift*
main road *139* *Hauptstraße*
mains services *66* *Versorgungsanschlüsse*
maintain *78* *warten*
maintenance *6* *Wartung*
major *47* *bedeutend, groß*
majority *72* *Mehrheit*
make friends *82* *Freunde finden*
make sure *107* *sich vergewissern, dafür sorgen*
make up *42* *bilden*
make use of *127* *benutzen*
management *41* *Geschäftsleitung, Verwaltung*
man-made *83* *künstlich, Kunst-*
mantle *124* *Ummantelung*
manufacturer *47* *Hersteller/in, Produzent/in*
manufacturing *42* *Produktion, Fertigung*
mark *28* *markieren, kennzeichnen*
marker *98* *Markierung*
market leader *47* *Marktführer*
marketplace *42* *Markt(platz)*
mass *124* *Masse*
material *6* *Angaben*
mathematical *10* *mathematisch*
matter *125* *Materie*
max. *25* *maximal*
maximum *61* *Höchst-*
meaning *25* *Bedeutung*
means of transport *106* *Verkehrsmittel*
meanwhile *99* *inzwischen*
measure *19* *messen;* *29* *Maßnahme*
measurement *94* *Abmessung, Maß*
mechanic *6* *Mechaniker/in*
mechanical *25* *mechanisch, technisch*
media hype *119* *Medienrummel*
melt *118* *schmelzen*
member *49* *Mitglied, Angehörige/r*
mercury *141* *Quecksilber*
message *10* *Nachricht*
metal cutters *16* *Seitenschneider*
metallic *73* *metallic*
micro-camera *131* *Mikrokamera*
microfilter *111* *Mikrofilter*
mile *73* *Meile*

pest 118 *Plage, Schädling*
petrol 84 *Benzin*
phase 114 *Phase*
phase out 141 *auslaufen lassen*
phenomenon 119 *Phänomen*
phone number 49 *Telefonnummer*
photocopy 97 *fotokopieren*
photograph 135 *fotografieren*
photographer 82 *Fotograf/in*
photovoltaic 147 *photoelektrisch*
physical 78 *körperlich, physisch*
pick out 143 *auswählen, heraussuchen*
picture in picture (PIP) 132 *Bild im Bild*
pie-chart 115 *Torten-, Kreisdiagramm*
pier 22 *Pier, Anlegestelle*
PIN code 138 *PIN-Nummer*
pincers 16 *Kneifzange*
pinning 85 *Nageln*
pipe 19 *Rohr*
pipe wrench 77 *Rohrzange*
pipeline 128 *Rohrleitung, Pipeline*
piston 84 *Kolben*
plane 92 *flach, eben*
planet 118 *Planet*
plant 6 *Werk, Betrieb*
plaster 102 *Gips, Verputz*
plastic 83 *Plastik, Kunststoff*
plate 101 *Platte*
platform ladder 31 *Trittleiter*
platinum 108 *Platin*
playback 132 *Playback*
pleasant 136 *angenehm*
pleased to meet you 70 *Es freut mich, Sie kennen zu lernen.*
pliers 16 *(Kombi-)Zange*
plodder 82 *Arbeitstier*
plughole 98 *Abfluss*
plumber 82 *Installateur/in, Klempner/in*
plumbing 25 *Klempnerarbeiten*
pm 29 *nachmittags*
point 91 *Punkt, Thema*
point in time 121 *Zeitpunkt*
point of view 91 *Standpunkt, Ansicht*
point out 100 *erläutern, erklären*
point to 71 *zeigen auf*
Poland 6 *Polen*
pole 131 *Pfahl, Mast*
police car 61 *Polizeiwagen*
police officer 65 *Polizeibeamter/-beamtin*
police station 36 *Polizeiwache, -revier*
policy 115 *Politik*
Polish 6 *Pole/Polin; polnisch*
poll 123 *Umfrage*
pollution 109 *(Umwelt-, Luft-)Verschmutzung*
polyester paint 84 *Polyesterfarbe*
polygon 93 *Vieleck, Polygon*
popular 72 *beliebt*
popularity 82 *Beliebtheit*
population 115 *Bevölkerung*
porous 109 *porös*
portable 108 *transportabel, tragbar*
Portugal 6 *Portugal*
Portuguese 6 *Portugiese/Portugiesin; portugiesisch*
position 134 *anbringen*
post 47 *aufgeben*
post office 18 *Postamt*
postal code 9 *Postleitzahl*
postcard 135 *Postkarte*
potentiometer 113 *Potentiometer*
power cable 44 *Stromkabel*
power line 78 *Stromkabel*
power point 44 *Elektroanschluss, Steckdose*
power supply 132 *Stromanschluss*
power unit 113 *Antriebsaggregat*
powerful 116 *mächtig, stark*
practicable 99 *brauchbar*
practical 82 *praktisch (veranlagt)*
practically 136 *praktisch*
pre-assembly 29 *Vormontage*
precious metal 107 *Edelmetall*
precise 51 *genau, präzise*
prepare 89 *vorbereiten*
present 121 *vorstellen, präsentieren*
presentation 100 *Vorstellung, Präsentation*
presenter 135 *Moderator/in*
preset 137 *voreingestellt*
pretty 142 *schön, nett*
price increase 145 *Preiserhöhung*
price 62 *Preis*
principle 99 *Grundsatz, Prinzip*
printer 148 *Drucker/in*
priority 25 *Vorrang, Priorität*
privacy 72 *Privatsphäre*
process 40 *Prozess, Ablauf;* **83** *bearbeiten*
produce 42 *produzieren, herstellen*
production planning 42 *Produktionsplanung*
progress 40 *Verfahren, Weg*
progressively 99 *zunehmend*
promise 29 *versprechen*
promising 123 *vielversprechend*
prompt 20 *Stichwort, Vorgabe*
prompt card 100 *Stichwortkarte*
properly 43 *richtig*
property 84 *Eigenschaft*
pros 72 *Für, Pro*
prospect 51 *Aussicht, Chance*
protect 44 *schützen*
proton exchange membrane 109 *Protonaustauschmembran*
proud 116 *stolz*
provide 114 *liefern*
pub 54 *Kneipe, Lokal*
public transport 128 *öffentliche Verkehrsmittel*
pulse 113 *pulsieren (lassen)*
pulsing 113 *Pulsieren*
pump 99 *pumpen*
pure 108 *rein*
purified 109 *gereinigt*
purple 81 *lila*
purpose 108 *Zweck*
push up 136 *hochtreiben*
put away 55 *weglegen*
put out 112 *löschen*
put up 139 *anbringen*
PVC 84 *PVC, Polyvinylchlorid*
pylon 78 *(Strom-)Mast*

Q

qualification 78 *Qualifikation, Fähigkeit*
qualified 26 *qualifiziert, ausgebildet*
qualify 79 *(sich) qualifizieren*
quality 30 *Qualität*
quality assurance 41 *Qualitätssicherung*
quantum dot 142 *Quantenpunkt*
quarter 58 *Viertel*
question tag 57 *Frageanhängsel*
questionnaire 80 *Fragebogen*
queue 60 *(Warte-)Schlange*

R

racing driver 82 *Rennfahrer/in*
radiation 101 *(Ab-)Strahlung, Abgabe*
radiator thermostat 144 *Heizkörperthermostat*
radical 115 *radikal*
radio frequency 136 *Rundfunkfrequenz*
radioactive 124 *radioaktiv*
railway station 131 *Bahnhof*
rain 65 *regnen*
rainbow 143 *Regenbogen*
rainwater 146 *Regenwasser*
range 99 *Bereich*
rapid 47 *schnell, rasch*
rarely 26 *selten*
rate 113 *Geschwindigkeit*
raw 83 *Roh(material)*
reach 34 *erreichen*
reach a peak 121 *einen Höchststand erreichen*
react 53 *reagieren*
reaction 52 *Reaktion*
realia 46 *Realien*
realise 62 *erkennen, feststellen*
reality 119 *Wirklichkeit, Realität*
rebuild 104 *wieder aufbauen*
receive 47 *erhalten*
receiver 137 *Empfänger*
recent 116 *jüngst, aktuell*
reception 41 *EmpfangRezeption*
receptionist 69 *Empfang(smitarbeiter/in)*
recessed 103 *eingelassen, umbaut*
recession 72 *Rezession*
recharge 110 *aufladen*
rechargeable 111 *aufladbar*

reckless *82 unbekümmert, rücksichtslos*
recognise *139 erkennen*
recommend *132 empfehlen*
recommendation *127 Empfehlung*
record *132 aufnehmen, -zeichnen*
recorder *44 Rekorder*
rectangle *92 Rechteck*
rectilinear *92 gradlinig*
recyclable *101 wieder verwertbar, recycelbar*
redesigned *116 neu konstruiert*
reduce *72 reduzieren, senken*
refer to *53 sprechen von, hinweisen auf*
reference *49 Referenz, (Arbeits-) Zeugnis, Empfehlungsschreiben*
reflex *93 überstumpf*
refresher *6 Auffrischung*
refreshments *89 Erfrischungen, Getränke*
refrigerant *99 Kühlflüssigkeit*
refrigerator *99 Kühlschrank*
refuel *114 auftanken*
refuse *139 (es) ablehnen, sich weigern*
region *99 Gegend*
register *68 zulassen*
registration *9 Zulassung, Registrierung*
regular *93 regelmäßig*
regularly *26 regelmäßig*
reject *43 ablehnen*
relate to *121 beziehen auf*
relationship *121 Beziehung, Verhältnis*
relatively *103 relativ, verhältnismäßig*
relax *60 (sich) entspannen*
relay *25 Relais*
release *99 freisetzen, -geben*
relevant *127 entsprechend*
reliability *117 Zuverlässigkeit*
reliable *117 zuverlässig*
remain *78 bleiben*
remote control *44 Fernbedienung*
removable *137 abnehmbar*
remove *19 entfernen*
renew *64 erneuern*
renewable *120 erneuerbar*
renovate *66 renovieren*
renovation *104 Renovierung*
rent *71 mieten; 75 Miete*
repair *23 reparieren*
repair work *23 Reparaturarbeit*
repeat *35 wiederholen*
replace *12 ersetzen*
replacement *127 Ersatz*
report *30 Bericht*
report back *122 Bericht erstatten, berichten über*
reported speech *149 indirekte Rede*
reprogram *138 neu programmieren*
request *23 Anfrage, Anforderung*
required *78 erforderlich*
requirement *78 Anforderung, Voraussetzung*
research and development *41 Forschung und Entwicklung*
reserve *97 reservieren, buchen*
reservoir *125 Stausee, -becken*
resident *66 Bewohner/in*
residual *102 Rest-, restlich*
respond *137 reagieren*
respondent *123 Befragte/r*
responsible *78 verantwortlich*
restate *72 neu formulieren, wiederholen*
result *61 Resultat, Ergebnis*
retailer *135 Einzelhändler/in*
retrain *6 umschulen*
return *62 Rück(reise); 99 zurückkehren*
reuse *137 wieder verwenden*
reverse *94 umgekehrt; 95 umkehren, vertauschen*
review *89 überprüfen*
revise *63 revidieren*
reward *47 Lohn, Vergütung*
rewrite *111 umschreiben*
Rhine *58 Rhein*
rhomboid *93 Rhomboid*
rich: the ~ *119 die Reichen*
ridge tent *94 Hauszelt*
right angle *93 rechter Winkel*
ring spanner *77 Ringschlüssel*
rise *118 (an)steigen; 121 Anstieg, Erhöhung*
risk-taker *82 risikobereite Person*
rivet *85 Niete*
riveter *85 Nieter/in*
road test *64 Straßentest*
roadworks *64 Straßenbauarbeiten*
robust *116 robust*
role *11 Rolle*
Romania *96 Rumänien*
roof *146 Dach*
rough *116 grob, hart*
roughly *48 ungefähr*
round *16 rund*
roundabout *35 Kreisverkehr, Kreisel*
route *62 Route, Strecke*
routine *23 Routine(arbeit)*
rowing boat *106 Ruderboot*
rubber *83 Gummi*
rubbish *71 Müll*
rule *16 Maß(stab), Lineal*
run down *107 sich entladen*
running cost *72 Unterhaltskosten, Betriebs(un)kosten*
rush-hour *61 Hauptverkehrszeit, Berufsverkehr*
Russia *6 Russland*
Russian *6 Russe/Russin; russisch*

S

sadly *142 traurigerweise*
safety *28 Sicherheit, Arbeitsschutz*
safety harness *28 Sicherheitsgurt*
safety measure *31 Sicherheitsmaßnahme*
safety net *31 Sicherheitsnetz*
sail *60 ablegen, den Hafen verlassen*
salary *47 Gehalt, Lohn*
sales assistant *136 Verkäufer/Verkäuferin*
sales tax *111 Mehrwerts-, Umsatzsteuer*
satisfied *104 zufrieden*
satnav *34 Navi(gationsgerät)*
save *33 sparen*
saving *141 Einsparung, Ersparnis*
scaffold tower *28 (Bau-)Gerüst*
scan *100 überfliegen*
scanner *136 Scanner, Lesegerät*
SCART jack *44 SCART-Stecker*
scooter *75 (Motor-)Roller*
score *10 Ergebnis, Punkt(estand)*
scoring *82 Punktestand, Auswertung*
Scotland *22 Schottland*
scream *61 heulen*
screed *101 Estrich*
screen *37 Leinwand*
screw *16 Schraube*
screw thread contact *141 Gewindekontakt*
screwdriver *16 Schraubenzieher, -dreher*
script *100 Manuskript*
seal *105 Siegel*
sealed *141 versiegelt, abgedichtet*
sealing *102 Versiegelung*
search *43 durchsuchen*
secondary *127 sekundär, Neben-*
secondly *108 zweitens*
section *85 Teil, Abschnitt*
sector *120 Sektor, Bereich*
secure *77 sichern*
security *40 Sicherheit*
security firm *130 Sicherheitsfirma, Wachdienst*
security guard *69 Sicherheitsbeamter/-in*
sedan *107 Limousine*
see you later *70 bis später*
segment *121 Teilstück, Segment*
seldom *26 selten*
self-confident *82 selbstbewusst*
self-employed *6 selbständig*
semicircle *94 Halbkreis*
sender *25 Absender*
sense of rotation *98 Drehrichtung*
sensible *82 vernünftig*
separate *77 trennen*
Serbia *6 Serbien*
Serbian *6 Serbe/Serbin; serbisch*
series *10 Reihe*
series production *143 Serienproduktion, -fertigung*
serious *123 ernst(haft)*
serve *43 servieren*

service *24 durchsehen, warten; 41 (Kunden-)Dienst*
session *91 Sitzung*
set off *65 in Gang setzen*
set up *139 aufstellen*
setting ring *98 Einstellring*
settle in *66 sich eingewöhnen, sich zurechtfinden*
severe *118 schwer, stark*
shape *19 formen; 92 Form*
-shaped *42 -förmig*
share *71 (die Wohnung mit jdm) teilen*
shared *66 gemeinsam*
sharp *116 scharf, stark*
sheet *19 (Metall-)Platte*
shelf *18 Regal(brett)*
shelving *40 Regale*
shift *23 Schicht*
shift allowance *23 Schichtzuschlag*
shine *65 scheinen*
shop assistant *136 Verkäufer/in*
shopkeeper *136 Ladeninhaber/in, -besitzer/in*
shoplifting *136 Ladendiebstahl*
shopping mall *37 Einkaufszentrum, -passage*
short cut *78 Abkürzung*
shortage *127 Mangel*
shoulder height *137 Schulterhöhe*
shout *100 schreien*
shower *23 duschen*
shy *82 schüchtern, scheu*
sided *92 -seitig*
signal *113 Signal, Zeichen*
significant *99 spürbar*
silver *73 Silber*
similar *67 ähnlich*
similarity *100 Ähnlichkeit*
since then *111 seitdem*
single *48 alleinstehend, unverheiratet; 62 einfach(e Fahrt)*
siren *61 Sirene*
site *Baustelle*
size *98 Größe*
sketch *18 Skizze*
skier *87 Skiläufer/in*
skilled *43 qualifiziert*
skills *49 Fertigkeiten*
skin *125 Haut*
skirting board *102 Fußleiste*
slider switch *98 Schiebeschalter*
slight *116 gering, leicht*
slimline TV *95 Flachbildschirm*
slope *87 (Ab-)Hang*
Slovak *6 Slowake/Slowakin; slowakisch*
Slovakia *6 Slowakei*
slump *72 Wirtschaftskrise*
smart *146 intelligent*
smell *44 duften, riechen*
smile *80 lächeln*
smoke *6 rauchen*
smooth *19 glätten*
snake *132 Schlange*
so far *42 bis jetzt*
so that *75 damit*
so-called *146 so genannt*
socket wrench *16 Steckschlüssel*
software engineer *82 Softwareentwickler/in*
soil *102 (Erd-)Boden*
solar *115 Sonnen-, Solar-*
solar cell *146 Solarzelle*
solar panel *148 Sonnenkollektor*
soldering iron *18 Lötkolben*
soldier *82 Soldat/in*
solid *92 fest, dreidimensional*
solid-fuel *99 Brennstoff-*
solution *99 Lösung*
solve *91 lösen*
someone else *82 jemand anders*
sort *42 sortieren*
sound *25 Klang; 68 klingen*
source *111 Quelle*
southern *99 Süd-, südlich*
Spain *6 Spanien*
Spanish *6 Spanier/in; spanisch*
spanner *16 Schraubenschlüssel*
spare part *33 Ersatzteil*
spark *25 Funke*
speaker *91 Redner/in, Person, die spricht*
special *72 speziell*
specialise in *135 sich spezialisieren auf*
species *118 Art*
specific *140 speziell, bestimmt*
specification *132 Angabe*
specified *104 angegeben*
speech bubble *6 Sprechblase*
speed *61 Geschwindigkeit*
speed limit *64 Geschwindigkeitsbegrenzung*
spend *56 verbringen*
spirit level *16 Wasserwaage*
spirits *134 Spirituosen*
sports car *116 Sportwagen*
spot check *96 Stichprobe*
spray *126 sprühen*
spy on *131 ausspionieren*
square *92 Quadrat*
stabilizer *28 Stabilisator*
stadium *96 Stadion*
staff *71 Personal, Mitarbeiter und Mitarbeiterinnen*
stainless steel *84 rostfreier Stahl*
standby switch *122 Ruhezustandsschalter*
staple *85 Heftklammer*
state *99 Zustand*
statement *24 Aussage*
statistics *130 Statistiken, statistische Angaben*
statue *100 Statue, Standbild*
steady *121 stetig, regel-, gleichmäßig*
steam *79 Dampf*
steel *83 Stahl*
steel plate *85 Stahlplatte*
steep *121 steil*
steering *64 Lenkung*
step *31 Stufe*
stomach *81 Magen*
stone *83 Stein*
stop and go *60 schleppender Verkehr*
storage *132 Speicherung, Archivierung*
store detective *136 Warenhausdetektiv/in*
store *23 (Material-)Lager; 42 lagern*
storm *118 Unwetter, Sturm*
stove *99 Ofen, Herd*
straight line *93 Gerade*
straight on *35 geradeaus*
strange *34 fremd*
stranger *38 Fremde/r*
street crime *135 Straßenkriminalität*
street plan *34 Straßenkarte*
stretched out *80 ausgestreckt*
strict *110 streng*
strip *101 Streifen*
structure *103 Struktur, Aufbau*
studio apartment *143 Einzimmerwohnung*
stuff *68 Kram*
subjected to *142 ausgesetzt*
submit *119 übersenden, aufgeben*
subsidise *122 subventionieren*
succeed in *139 es erreichen*
successful *47 erfolgreich*
such as *51 wie (zum Beispiel)*
suddenly *30 plötzlich*
sugar cube *94 Würfelzucker*
suggest *122 vorschlagen*
suggestion *122 Vorschlag*
suitable *24 passend*
suitcase *40 Koffer*
sunlight *44 Sonnenlicht*
sunroof *38 Schiebedach*
superior *110 überlegen*
supertram *37 Großraum-Straßenbahn*
supplement *99 ergänzen*
supplier *42 Lieferant/in, Zulieferer*
supply *49 liefern; 78 Lieferung, Zufuhr*
support *28 ab-, unterstützen*
suppose *43 glauben, annehmen*
supposed to, be ~ *142 sollen*
surface *19 (Ober-)Fläche*
surname *7 Nach-, Familienname*
surprised *43 überrascht*
surprising *116 überraschend, erstaunlich*
surround *28 umgeben*
surveillance camera *40 Überwachungskamera*
surveillance equipment *137 Überwachungsgerät(e)*
survey *98 (Meinungs-)Umfrage*

unrest *127 Unruhe(n)*
unsatisfactory *49 unbefriedigend, ungenügend*
unusual *51 ungewöhnlich, unüblich*
up to *58 bis zu*
update *89 aktualisieren, auf den neuesten Stand bringen*
upmarket *131 oberes Marktsegment*
upon request *49 auf Anfrage*
upset *82 verärgern, aufregen*
upside *88 Vorzug*
urgent *25 dringend, eilig*
used *73 gebraucht*
useless *138 nutzlos, unbrauchbar*
user manual *117 Benutzerhandbuch*

V

vacation *47 Urlaub, Ferien*
value *72 schätzen, Wert legen auf*
van *20 Transporter*
vaporise *126 zerstäuben*
vapour *126 Dampf*
variable resistor *113 regelbarer Widerstand*
varied *77 abwechslungsreich*
variety *51 Abwechslung*
various *37 verschieden*
vary *100 variieren, ändern*
VCR *132 Videorekorder*
vehicle *38 Fahrzeug*
vending machine *134 Verkaufsautomat*
ventilation *147 Lüftung, Belüftung*
venue *90 (Tagungs-)Ort*
versatile *100 vielseitig*
vertical *141 senkrecht, vertikal*
via *59 über*
vice *17 Schraubstock*
vice versa *98 (und) umgekehrt*
Victorian *66 viktorianisch*
video graphics array (VGA) *132 Videofilmaufzeichnung*
video monitor *43 Bildschirm*
Vienna *14 Wien*
view *127 Ansicht, Standpunkt*
viewer *132 Überwachung*
viewing screen *132 Bildschirm*
violence *138 Gewalt*
visible *124 sichtbar*
vivid *142 lebhaft*
voice *100 Stimme*
volatile *107 flüchtig*
volcanic *124 vulkanisch*
volcano *124 Vulkan*
voltage controlled amplifier (VCA) *132 VCA-Verstärker*
voltage *77 Spannung*
voltage tester *77 Spannungsprüfer*
voltaic *146 voltaisch*
volume *100 Lautstärke*
vs. *127 gegen(über)*

W

wages *87 Löhne*
walkway *42 Gang, Fußweg*
warn *64 warnen, darauf hinweisen*
warning light *61 Warnlicht*
warning sign *28 Warnzeichen, -schild*
washbasin *98 Waschbecken*
washing machine *78 Waschmaschine*
washing-up *43 Abwasch, Spülen*
waste *111 vergeuden, verschwenden*
watch out! *97 Vorsicht! Pass auf!*
water level *118 Wasserspiegel*
wax *83 Wachs*
weakness *53 Schwäche*
weekday *66 Wochentag, Werktag*
weekly *105 wöchentlich*
weight-carrier *114 Lastenträger*
welcome to *68 willkommen bei*
welded *77 geschweißt*
welder *76 Schweißer/in*
welding torch *77 Schweißbrenner*
well *124 Brunnen*
what about *51 was ist mit?*
what else *29 was sonst (noch)*
what kind of *80 welche Art von*
whatever *139 was auch immer*
wheel *94 Rad*
whereas *110 während*
wherever *127 wo (immer)*
whether *130 ob*
white *81 weiß*
whitish *143 weißlich*
whole *73 ganz*
wide *80 weit*
width *92 Breite*
willing *43 bereit(willig)*
win *128 gewinnen*
wind farm *115 Windkraftanlage*
wind turbine *146 Windturbine*
window *33 Fenster*
windscreen wiper *64 Scheibenwischer*
winner *119 Gewinner/in*
wire *16 Draht*
wire strippers *17 Abisolierzange*
wireless *131 drahtlos, Funk-*
wiring *78 Verkabelung, Leitungen*
wise *124 weise, klug*
wish *82 wünschen*
with great difficulty *135 unter großen Schwierigkeiten*
witness *65 Zeuge/Zeugin*
wonder *42 sich fragen; 146 Wunder*
wonderful *38 wunderbar*
wooden *104 hölzern, Holz-*
wool *83 Wolle*
wordfield *58 Wortfeld*
work experience *49 Berufs-, Arbeitserfahrung, Praktikum*
work out *11 ausrechnen, kalkulieren*
workbench *12 Werkbank*
working day *23 Arbeitstag*
workplace *51 Arbeitsplatz*
worksheet *33 Arbeitsbogen*
workshop *41 Werkstatt, Werkshalle*
worldwide *138 weltweit*
worried *115 besorgt*
worrier *82 sorgenvoller Mensch*
worry *64 sich Sorgen machen*
worrying *118 beunruhigend*
worthwhile *77 lohnend*
write in full *66 ausschreiben*
writing *73 Schreiben*
wrongdoer *131 Missetäter/in*

X, Y, Z

X-ray *7 Röntgenstrahl*
yard *42 Hof*
yellow *31 gelb*
yellowish-white *143 gelblich-weiß*
you're welcome *36 gern geschehen*
Yours sincerely *48 Mit freundlichen Grüßen*
youth hostel *36 Jugendherberge*
zone *103 Zone, Gebiet*

Quellennachweis

RF photos
Alamy: S. 7 A.Skelley, S. 23/2 P. Hill, S. 76/6 D. Wassermann, S. 79/2 Photodisc, S. 79/3 G. Costanzo,
S. 79/4 ImageSource, S. 130/6 G. Cali, S. 131/1, S. 141/1 F. Szelepcsemy, S. 141/2 Digifoto Gamma;
Corbis: S. 6/1, 4, S. 76/3 N. Amonn, S. 78/2 Photolibrary;
Istockphoto: S. 6/2, S. 67/1;
V. Maly: S. 46/6;
Masterfile: S. 30;
R. Nadolny: S. 95/5, 6;
Shutterstock: S. 6/3, 5, 6, S. 23/1, S. 49, S. 67/2–6, S. 95/1–4, S. 106/1–8, S. 107/1, S. 108

RM photos
Alamy: S. 58/1 J. Tack, S. 58/2 brinkstock, S. 58/3 qaphotos.com, S. 76/1 stockphoto/Nir Alon, S. 76/4 D. Green,
S. 76/5 M. Jenkinson, S. 76/7 RightImage, S. 76/9 W. Dietrich, S. 78/1 K. Hildebrand, S. 78/3 M. Jenkinson,
S. 79/1 blickwinkel, S. 83/1 Mark&Audrey Gibson, S. 83/2–3 VarioImages/U. Baumgarten, S. 130/1 D. Pearson,
S. 130/2 Caro/Bastian, S. 130/4 VarioImages/J. Wolter, S. 130/7 E. Moss, S. 131/2, S. 144/1 S. Klein,
S. 144/2 artpartner-images.com, S. 144/3;
Fotofinder: S. 10 VarioImages, S. 23/3 Visum/S. Döring, S. 46/4 Ostkreuz/M. Trippel, S. 46/5 VarioImages,
S. 46/7 Keystone/J. Zick, S. 46/8 Image Source, S. 50 Arturimages/C. Richters, S. 71 F1 online/J. Lecher, S. 76/2 G. Alabiso,
S. 76/8 Presse Foto Design/K. Fuchs, S. 76/10 BilderBox, S. 110/1 Bildagentur Fnoxx/A. Hettrich, S. 112 Dhote/Andia,
S. 118/1 Argum/T. Einberger, S. 118/2 Wildlife, S. 118/5–6 M. Weber, S. 130/3 fotohardy, S. 130/8 A1PIX;
Picture Alliance: S. 107/3 dpa-Report, S. 118/3–4

Für die freundliche Unterstützung danken wir folgenden Unternehmen:
S. 60/2 TrafficLinq.nl;
S. 107/3 BMW;
S. 110 General Motors Corporation;
S. 132 Luxor Direct on behalf of CloseoutCCTV;
S. 136 ezy-tag.de;
S. 140/1 MACtac Deutschland GmbH;
S. 140/2 CV-Archiv;
S. 140/4 Lin-Wang Wang;
S. 140/6 Optics.org/IOP Publishing;
S. 140/7 General Motors Corporation;
S. 143 Collegiate Images/Vanderbilt University;
S. 144/4 Utica Boilers/ECR International;
S. 146/1 Premier Electric Blinds;
S. 146/4 AP Images/Koji Sasahara;
S. 146/6 Rehau AG + Co./Viessmann Werke GmbH & Co KG